MW00560270

THE GREENING OF ASIA

THE GREENING

OF ASIA

The Business Case for Solving Asia's Environmental Emergency

MARK L. CLIFFORD

Columbia University Press
Publishers Since 1893
New York Chichester, West Sussex

Library of Congress Cataloging-in-Publication Data

Clifford, Mark, 1957–
The greening of Asia : the business case for solving Asia's environmental emergency / Mark L. Clifford.
pages cm. — (Columbia business school publishing)
Includes bibliographical references and index.
ISBN 978-0-231-16608-9 (cloth : alk. paper) —
ISBN 978-0-231-53920-3 (ebook)
1. Sustainable development—Asia. 2. Environmental degradation—Asia. I. Title.

HC415.E5C53 2015
338.95'07—dc23

2014029476

Columbia University Press books are printed on permanent and durable acid-free paper.
This book is printed on paper with recycled content.
Printed in the United States of America

c 10 9 8 7 6 5 4 3 2 1
p 10 9 8 7 6 5 4 3 2 1

COVER DESIGN: Noah Arlow

CONTENTS

PREFACE

The beginning of what became a seven-year search for Asia's green shoots reaches back to Mumbai in May 2007. I had just joined the Asia Business Council after 25 years in journalism, and while in India I met Jamshyd Godrej, whose Godrej & Boyce group is one of India's oldest and best-regarded companies. A project that Jamshyd had developed in Hyderabad had recently been ranked as the world's most energy-efficient building of its type. Jamshyd and his team's work symbolized a new phase to the rise of Asia, one that moved beyond low-end manufacturing.

After the publication of an Asia Business Council book on green buildings later that year, I began to look more broadly at what Asian business was doing to meet the region's environmental emergency. The more I looked, the more evidence I found of Asian companies engaged in large-scale business activities designed both to profit from and to help solve Asia's environmental challenges.

At a conference in Hong Kong at the end of 2007, Andrew Brandler, the CEO of CLP Holdings, announced the company's promise to reduce the carbon intensity of its electricity by more than seventy-five percent by 2050. This was a globally unparalleled pledge by a major utility. Subsequent conversations with Andrew provided the catalyst to embark on this book; Andrew and others at CLP were a major source of support.

In April 2010, I spent a fascinating day with Dong S. Kim of the SK Group, one of South Korea's *chaebol*, a group whose annual sales total more than $100 billion. At the SK Energy Research Institute outside of Seoul, Kim oversaw an annual budget of $250 million that included spending on a number of clean energy projects, including one to produce plastics from carbon dioxide emissions. It was a sustainability-focused research budget that few companies anywhere could match.

A few weeks later, back in Mumbai, Anand Mahindra—a graduate of both Harvard College and Harvard Business School—told me that focusing on sustainability would create a new wave of growth opportunities for his company, Mahindra & Mahindra. Anand recounted a meeting he had just attended with a group of Indian government officials and an NGO representative. Everyone in the group seemed to think that environmental sustainability could only come at the cost of economic growth. "I said it reminded me of this time in the 1980s where people used to see quality and cost as antithetical. Americans were scratching their heads and saying 'how do the Japanese get ahead of us in quality?' Then some Americans went and had a eureka moment and they realized that better quality lowers cost. The world is waiting for a eureka moment in sustainability. Sustainability is an opportunity for growth. If you can make that shift, you are into a whole new phase of growth."

In the chapters to come, I'll look at the environmental challenges that companies face, and will profile a number of Asian businesses that are taking innovative measures—in areas from solar and wind power to green buildings and electric cars to water services and sustainable tropical forestry—in response. These profiles show how private firms are using technology, money and, above all, employee ingenuity to begin solving Asia's environmental challenges. These activities are not greenwashing, philanthropy, or "corporate social responsibility," but hard-headed business reponses to opportunity born out of crisis. The final section of the book is an appendix (Companies to Watch) comprising brief profiles of key environmental businesses of some representative companies, mostly Asian.

I don't want to oversell how widespread this transition is, or how fast it's happening, but visits to factories and farms throughout Asia convinced me that this transition is for real. It is happening because there are business opportunities.

But businesses can't act alone. Many Asian companies are ready, willing and able to act to help solve the region's environmental problems—but governments do not know how capable these companies are, and they do not understand the powerful role that private companies can play in solving social problems. The state needs to set the rules of the game, to aid and incentivize greener business. What sorts of rules are needed? That's one of the major questions this book will explore. From higher prices for scarce resources and an end to fossil fuel subsidies, to better engagement with grassroots groups and civil society, government can play a powerful role in addressing one of the most important struggles of our century—how to live on a planet whose resources we are consuming too fast. We can live well, and we can continue to raise the quality of life for the billions of people who do not have light or running water in their homes as well as for those city-dwellers whose hopes of a better future are endangered by pollution and shoddy infrastructure. Asia has focused on top-line GDP growth, which is politically popular. And certainly, growth matters. But the quality of growth matters now as never before.

This book is not encyclopedic—crucial topics such as sustainable fishing and the promise of nuclear power are covered only in passing. And there are businesses I had to leave out as well—including the SK Group and Mahindra & Mahindra. But what I do hope you'll come away with is a sense of possibility and promise. Sustainable growth is possible, and businesses—so often painted as villains, and sometimes rightly so—must lead the way. I am indebted to many people who shared their time and expertise with me. Some of them are quoted in this book, but most are not. The list below of those who helped me is partial, but acknowledges my principal intellectual debts.

Above all, I want to thank the trustees of the Asia Business Council who gave me the freedom to pursue this project as I wished. I especially want to thank Council Chairs Narayana Murthy, Qin Xiao, Marjorie Yang and Jaime Zobel de Ayala.

My agent Leah Spiro, whom I met when we were at *BusinessWeek*, was key to helping me marshal my material. At Columbia University Press, thank you to Myles Thompson for understanding this project and for Bridget Flannery-McCoy for her magical work with the manuscript. If Myles and Bridget are any indication, the book industry has a bright future ahead of it; they and their colleagues at Columbia, notably Stephen Wesley, share a love of words and ideas that exemplifies what is best about publishing. Ben Kolstad expertly shepherded the book through the production process. Many people were extraordinarily generous with their time. I mention only the most important here; I also particularly want to acknowledge the logistical efforts of many staff members who are not named but facilitated my trips and interviews.

At ANJ, George Tahija and, in Belitung, Philip Liu, and Joseph Gomez; at BYD, Wang Chuanfu, Michael He, and Sherry Li; at CLP, in addition to Andrew Brandler, Michael Kadoorie, Richard Lancaster, Peter Greenwood, Jane Lau, Jeanne Ng, and Dorothy Chan; at Esquel, in addition to Marjorie Yang, Agnes Cheng and Dodie Hung; at Godrej & Boyce, in addition to Jamshyd Godrej, Anup Mathew and Rumi P. Engineer; at Goldwind, Wu Gang and Kathryn Tsibulsky; at Gale International, Stan Gale and Scott Summers; at Great Giant Pineapple, Husodo Angkosubroto and Ruslan Krisno; at Hang Lung Properties, Ronnie Chan; at Hyflux, Olivia Lum; at Infosys, Kris Gopalakrishnan; at Lafarge, Bruno Lafont, Lionel Bourbon, and Koul Nilesh; at Mahindra & Mahindra, in addition to Anand Mahindra, Rajeev Dubey and Beroz Gazdar; at REVA Mahindra, Chetan Maini and R. Chandramouli; at Manila Water, Gerry Ablaza and Carla Kim; at McKinsey, Dominic Barton, Adam Schwarz, and Jonathan Woetzel; at the MTR, Glenn Frommer; at Orix, Yoshihiko Miyauchi and Atsushi Murakami; at SCG (Siam Cement) Kan Trakulhoon and Cholathorn Dumrongsak; at Schneider, Jean-Pascale Tricoire; at Singbridge, Lim Chee Onn and Koh Kheng Hwa; at the Singapore PUB, Chew Men Leong; at the SK group, Chey Tae-won and S. K. Dong; at Swire, Philippe Lacamp; at Toyota, Bernard O'Connor and Akiko Machimoto.

A presentation of a draft of the introductory chapter in November 2013 at the Hong Kong Forum allowed me to sharpen many of the book's ideas. Christine Loh and Edgar Cheng were excellent interlocutors, but that event was only one of the many ways in which both Christine and Edgar have helped me over the years; thanks also to Elaine Pickering of the Forum secretariat as well as all those who participated in the session.

Others who generously shared their expertise with me included Andrew Affleck, Toshio Arima, Charles Bai, Nancy Bowen, Tim Dattels, Liz Economy, Brooks Entwistle, Barbara Finamore, Rachel Fleishman, Patricia Gallardo, Julian Goh, Kristine Johnson, Calvin Lau, Dirk Long, Jim Maguire, Steve Markscheid, Michael McElroy, Tak Niinami, Doug Ogden, Michael Polsky, Alan Rosling, Orville Schell, Nigel Sizer, Alex Tancock, Edith Terry, Xu Yuan, Jerry Yan, Charles Yonts, and Daniel Zuellig.

Thanks also to my anonymous readers at Columbia University Press, whose comments helped me strengthen the book.

I owe a special debt to the Council's Adjunct Fellow Jill Baker, whose enduring commitment to a project that took much longer than envisioned when we first met in early 2012 has been invaluable. The solar and wind chapters in particular benefitted from Jill's financial expertise but her work on the entire manuscript has made this an immeasurably richer book. Program Director Janet Pau as well as the Council's Princeton in Asia Fellows Kaishi Lee, Kari Wilhelm, Tom London, Matt Garland, Ryan Brooks, and Jack Maher all provided research help. I also want to thank others on the Council's staff: Administrative Director Winnie Wu as well as Bonnie Chang, Alex Zhang, Jennifer Wei, Philip Hui, and Moneta Chon. In Whistler, Scott and Helen Carrell kept me fed, fit and happy while I wrote in the mountains. Finally, thanks to our children, Anya and Ted, and, above all, my wife, Melissa Brown, whose knowledge of many of the issues discussed in this book is exceeded only by her patience in enduring an even more fraught book-writing process than usual is more than I could have asked for.

A final note: Some of those quoted in the book are members of the Asia Business Council; many are not. This book was written independently and the opinions are mine. So too are any errors, whether of commission or omission. I hope that this work will stimulate debate, inspire confidence that there is a way forward, and lead to faster action to solve one of the defining challenges of our age.

Hong Kong, October 2014

THE GREENING OF ASIA

Introduction:
Green Shoots Under Soot-Stained Skies

> The difficulty lies, not in the new ideas, but in escaping from the old ones.
>
> —JOHN MAYNARD KEYNES

Beijing's air is "crazy bad," according to the U.S. Embassy: choking pollution regularly smothers the capital, reducing visibility to near zero, grounding planes, snarling traffic, and forcing city dwellers to don protective face masks while outside. A widely used air quality index, which in the United States rarely goes above 100 and exceeds 300 only during forest fires and other extreme events, approached the 1,000 level in Beijing in early 2013.

The effect, says a Chinese researcher, is to blot out the sun as effectively as a nuclear winter. Office workers in the capital's skyscrapers cannot see the streets below, as a bitter, blinding pall settles over a city that hosted the 2008 "Green Olympics." Beijingers call it "air-pocalypse" or "air-mageddon," and they have become increasingly vocal about their frustration. "I especially want to know if the party secretary or the mayor are in Beijing these days," a senior editor at *People's Daily* wrote on his blog during record smog in January 2013. "If so, how do they guarantee they can breathe safely in Beijing?"[1]

Two thousand miles to the south, in Nga Pi Chaung, Myanmar, some forty seventh graders crammed into a small classroom are learning about climate change. Their small village clings to the belly of Myanmar's Ayeyarwady Delta and is without electricity or running water. It is also a place with no litter—its 821 inhabitants are subsistence farmers on the margins of the cash economy, too poor even to generate trash. In 2008, the delta was shredded by Cyclone Nargis, and an estimated 138,000 people were killed. And although no one knows definitively if Cyclone Nargis was caused by climate change, what's clear is that big storms are coming more often and are more violent when they hit.[2] For students in this village, for their parents, and for tens of millions of people in Myanmar and hundreds of millions of people throughout Asia, this pattern of more extreme weather threatens their already difficult lives. In some cases, it threatens the nations themselves. Bangladesh, Myanmar's neighbor to the west, has a population of 160 million people, and most of them live fewer than ten feet above sea level. Even a slight rise in the sea level could put the country basically underwater. The Maldives faces a similar fate: "If things go business-as-usual, we will not live, we will die," Maldives President Mohamed Nasheed told the United Nations in 2009. "Our country will not exist."

Asia's richer nations are not immune to environmental emergencies. More than a thousand miles south of the Ayeyarwady Delta, almost on the equator, is the wealthy city-state of Singapore. With average incomes higher than those in the United States, it prides itself as a city in a garden, one whose commitment to the environment sets it apart from the region. But in June 2013, the tree-filled city was blanketed by choking haze from fires in neighboring Indonesia. These fires, most of which were deliberately set as a quick way of clearing land for palm plantations, forced Singaporeans indoors and spurred the formation of a high-powered Haze Inter-Ministerial Committee, chaired by the defense minister. "Singapore . . . is being suffocated," lamented former Prime Minister Goh Chok Tong. Current Prime Minister Lee Hsien Loong was blunt when it came to assigning blame: the fires occurred because of "illegal burning by errant companies" in Indonesia.[3] The burning of Indonesian forests is doubly damaging, robbing the planet both of its biodiversity and of one of its most important carbon sinks.

Asia is approaching a moment of systemic—in some cases, existential—crisis. How Asian countries react to the environmental challenges of pollution, resource shortages, and climate change will determine whether the region will continue along its unmatched path of growth or descend into an increasingly unlivable dystopia. Its response will be unlike that of the United States or Europe; Asia has more people, more governments, more diversity in terms of wealth and poverty, and a broader range of natural resource endowments. And although this diversity makes the path to solutions more complex, it has also positioned Asia as a test-bed for hundreds of experiments, both in businesses and in government policies. Governments play a critical role in setting and enforcing rules, but it is companies that bring government policies to life in the real world. Individual firms have the resources—the money, the people, and the technological know-how—to most effectively implement change. Asia already has countless companies that are, in ways large and small, making the transition to a greener, less resource-intensive, lower-carbon world. The chapters that follow profile some of these companies and explore the essential role that business can and must play in helping ensure that Asia chooses a path to ensure long-term prosperity. But in order to understand the impact that companies can have, it's important to take a closer look at the way business, policy, and civil society interact and how these three forces can align to create change.

~

Asia is the greatest economic success story in human history. Hundreds of millions of people have worked their way out of poverty. In 1981, three out of four people in East Asia lived in extreme poverty, surviving on less than $1.25 a day; by 2008, it was only one out of seven. The economies of Japan, Korea, Taiwan, Singapore, Hong Kong, and, of course, China have grown at faster rates and for a longer stretch of time than the world has ever seen. A clutch of other countries, including Indonesia, Malaysia, Thailand, and, more recently, India and Vietnam, have also grown at unusually high rates. Often termed the "Asian miracle," this growth is in fact the product of hard work, increased education, and a pro-market bias buttressed by supportive government policies and a liberal global trade environment. Asia's newfound prosperity is the best evidence that education, globalization, and good government policies can transform human lives, opening up new opportunities to people whose parents and grandparents struggled to survive, their lives bounded by poverty, hunger, and disease.

China is the giant among these growing giants; since 1981, 660 million Chinese have moved out of the ranks of the extremely poor. This is in part due to the country's historically unprecedented average annual growth rate of almost 10 percent from the time economic reforms began in 1978 until 2013. By comparison, Great Britain, during the height of its Industrial Revolution, grew just over 2 percent a year. China has doubled the size of its economy roughly every seven years, and it is now, and this is an incredible number to consider, about *thirty times* larger than it was in 1978, when economic reform began.[4]

Smog shrouds Beijing's Fourth Ring Road. Photo credit: David McIntyre

But this growth without parallel has led to pollution, resource scarcity, and environmental woes without precedent: most of Asia remains stuck in the "get dirty, get rich" phase of development, and many of Asia's 4.4 billion people—six out of every ten people in the world—live in countries where economic expansion is putting extraordinary and often unsustainable pressures on the natural environment. The world's ten most polluted cities are in the region.[5] Clusters of so-called cancer villages are a stain on an otherwise enviable record in improved life expectancy. Southeast Asia's tropical forests are being felled to make way for palm oil plantations. The water in many Asian rivers is too toxic even to touch. Most troubling of all, air pollution in China alone contributed to more than 1.2 million premature deaths in 2010—an annual death toll larger than the population of San Diego or Dallas. This is part of the price Asia has paid for its economic progress.[6]

Environmental issues don't respect national boundaries, and problems that originate in one country may settle in another. The haze from Indonesian fires over Singapore is just one example. China's neighbors to the east, Korea and Japan, suffer from China's air pollution as well as the massive seasonal sandstorms that result from the deterioration and desertification of the land west of Beijing. The fallout from Asia even affects the United States: California researchers detected record levels of radioactive sulphur following the March 2011 Fukushima nuclear disaster; mercury pollution from China reaches California on an ongoing basis, carried on air currents high above the Pacific Ocean; and an Asia-grown soup of ozone, carbon monoxide, and particulates drifting over the West Coast of the United States may be contributing to more serious storms in California and the Pacific Northwest.[7]

Asia's environmental woes are likely to get worse before they get better. In part, this is due to the happy circumstance that the lives of the Asian poor are likely to continue improving. The Asian Development Bank says that 628 million people in Asia are still without electricity; this means that even in the countries participating in the world's great economic success story, one out of every five people is still living in a house without even a single electric lightbulb.[8] As the poor continue to rise out of poverty and as the population as a whole continues to grow, there will be an extraordinary strain on resources—water, food, fuel—and there is likely to be an increase in carbon emissions as well.

China is already the world's largest emitter of greenhouse gases; the country's 1.4 billion people are responsible for over one quarter of global greenhouse gas (GHG) emissions. Indonesia, with its rapid deforestation,

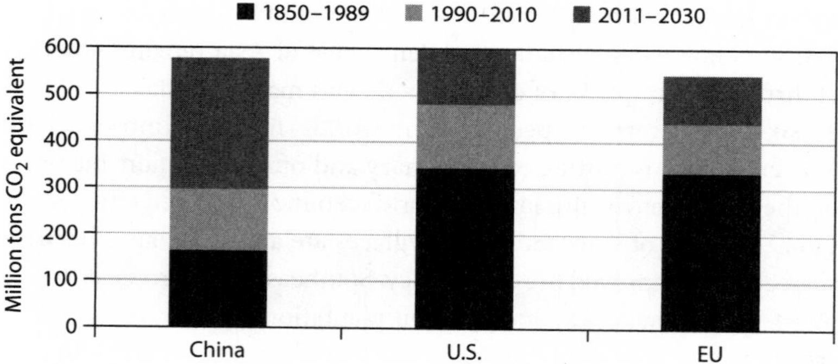

Greenhouse Gases
By 2030, China is projected to surpass Europe and almost catch up with the United States in total greenhouse gas emissions since 1850. Source: PBL Netherlands Environmental Assessment Agency

by some measures now ranks third in total GHG emissions. India is at a similar level.[9] These three countries—China, India and Indonesia—are, along with the United States and Russia, responsible for half of all annual GHG emissions. Any attempt to slow the effects of global warming has to include these Asian giants.

Asia's big three are not rich countries. But there is a new model of economic development, one that relies on a mixture of increased energy efficiency—especially for buildings and vehicles—coupled with a much higher reliance on renewable energy and an end to fossil fuel subsidies. Asian companies are ready for the challenge. Asian governments need to put the rules in place, quickly, to accelerate this transformation.

What can't go on forever, to paraphrase American economist Herbert Stein, will stop. Just as the frenetic pace of Asian economic growth is beginning to slow, Asia's environmental degradation, especially water and air pollution, is approaching a point when the political, social, and economic costs of *not* acting will be too expensive. Coupled with the challenge of reducing pollution will be the increasing impact of climate change, which is likely to exacerbate these already difficult issues. The United States, with its relatively low population spread over a vast land area, its wealth of resources, and its high income, can to a large degree wall itself off from the effects of climate change. Storms like 2012's Hurricane Sandy may cause tens of billions of dollars of damage, but they do not, like Asia's major typhoons, kill thousands of people.[10] Asia's population density means its people are unusually vulnerable to environmental problems, including clean water shortages, air pollution, dismal mega-cities, and the increasingly unpredictable effects of climate change.

When looked at from this angle, a more environmentally resilient Asia looks less like a matter of choice and more like a matter of survival. An optimistic reading of the current situation is that increasingly worse environmental problems will spur change. This has historically been the case in other countries, like Britain and the United States, which were forced into action as a result of disasters that occurred after a period of rapid industrialization. Britain introduced landmark clean air legislation following the toxic fog of December 1952, which killed at least four thousand and perhaps as many as twelve thousand people. A June 1969 fire on Ohio's Cuyahoga River—"a river that oozes rather than flows," *Time* wrote—catalyzed the environmental movement in the United States; the Environmental Protection Agency was inaugurated eighteen months later.[11]

The same pattern occurred in Japan, which has long been the Asian pacesetter when it comes to environmental awareness. It has seen its share of headline-grabbing pollution disasters, notably mercury poisoning in Minamata and choking air pollution in Yokkaichi, both of which have the dubious distinction of having diseases named after them (Minamata disease and Yokkaichi asthma). Those illnesses, especially the Minamata mercury poisoning, led to more environmental awareness. For Japan, as for Britain and the United States, the public played a major role in forcing the government to mandate change following these disasters.

Japan's energy and environmental consciousness started at the popular level with anger at Minamata and Yokkaichi, but the 1973 Arab oil embargo and a second oil shock at the end of the decade focused industry and government attention on energy price and availability. As an island nation with few natural fuel resources, Japan began to embrace the idea of energy efficiency.

Today, Japan has some of the best environmental protection and some of the most sweeping energy-conservation measures in the world. It is also among the world's more energy-efficient developed economies. From steel mills to autos, the Japanese have learned to do more with less. Crucially, this involved both conservation in the negative sense—for the American version of this, think of Jimmy Carter with his sweater in the White House, preaching about the need to save energy—and efficiency in the positive sense. Their fuel-efficient cars, for instance, allowed Japanese manufacturers to make inroads against their American rivals, leading three decades later to Toyota's line of Prius and other models of hybrid cars; more than six million of these vehicles have been sold since their introduction in 1997.

Japan is poised to become even more efficient: the devastating March 2011 earthquake and tsunami that resulted in a meltdown at the Fukushima nuclear plant led its government to promise, initially, to phase out nuclear energy by 2040. Although a new government has disavowed the no-nuclear promise, the country is already taking steps to dramatically increase the role of renewable energy, especially solar power. In 2012, it introduced extremely generous prices for electricity produced by solar power as a way of boosting the industry.[12]

Japan serves as an excellent example of how the triple forces of public pressure, political support, and business-led innovation can help reshape a country's approach to energy and the environment. Indeed, the presence of these three interrelated forces seems crucial if other Asian countries are to follow Japan's green path. And all three forces can be tracked in the country in which change matters most—China.

China's size and speed of growth make it the key variable in the Asian environmental equation. Public outcry has begun: the unprecedented public anger in Beijing at the "air-pocalypse" of January 2013 may herald a shift in the country similar to what Japan went through after Minamata and Yokkaichi. The "air-pocalypse" followed numerous demonstrations in 2011 and 2012 that hinted at a newfound unwillingness by ordinary Chinese to accept environmental degradation—and the health costs that come along with pollution—as the price for economic growth. When Xi Jinping became the country's new president in March 2013, environmental issues were regularly cited as one of his most pressing tasks.

But even prior to this public pressure, the Chinese government seems to have realized the necessity of moving to a less-polluting, lower energy-intensity, and lower-carbon economy. The 2006–10 Five-Year Plan period—the first with an energy-intensity target—saw the country reduce its reported energy consumed per unit of economic output by an impressive 19 percent, close to the plan's 20 percent target. A further energy-intensity reduction of 16 percent is targeted for the 2011–15 Five-Year Plan period. Impressive as these targets sound, overall growth was substantially higher than the efficiency gains—the economy expanded by 70 percent in the 2006–11 period—so the growth in both overall pollutants and CO_2 emissions was still substantial. Moreover, China is still amazingly inefficient in its use of energy when compared to other countries, needing, for instance, about three-and-a-half times as much energy per unit of economic output as the United States.[13]

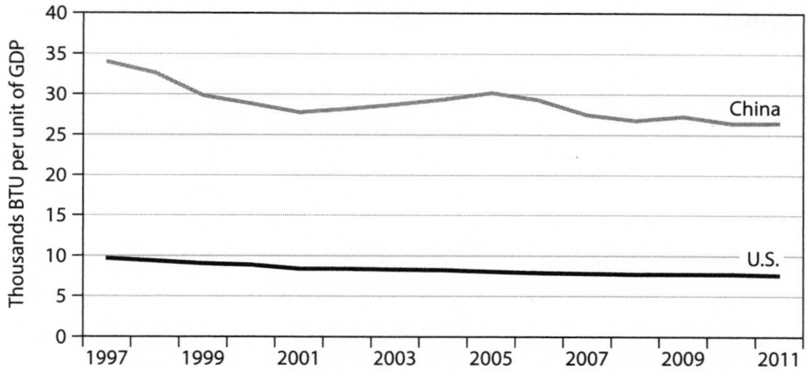

Energy Efficiency in China and the United States
China's energy-intensive economy has the potential for dramatic improvement.
Source: U.S. Energy Information Administration

Although far behind some of its peers at this point, China is investing more than the United States in renewable energy, even though its economy remains only a bit more than half the size of the U.S. economy. According to Bloomberg New Energy Finance, China invested a total of $125 billion in clean energy in 2012 and 2013, compared with the United States's $101 billion. For several years, China has been the world's largest manufacturer of wind and solar power equipment. Now, it is on the way to becoming the largest consumer—thanks in part to the fact that the country's planners have set a target of 15 percent of its energy from renewable sources by 2020.[14] In 2009, it became the largest market for wind turbines, a position it has, with the exception of 2012, occupied ever since. In 2013, it installed more than 12 GW of solar power—before 2013, no country had ever installed more than 8 GW in a single year. Also in 2013, it for the first time spent more than the United States on building out its so-called smart grid, designed to make electricity generation and distribution more energy-efficient.[15]

The 2011–15 Five-Year Plan also aims to double the environment-related market.[16] In 2011, Deborah Seligsohn, an expert on the Chinese environment who at the time was based in Beijing for the World Resources Institute, told a U.S. House subcommittee: "China missed the industrial revolution, it was late to the IT revolution, and they see this new clean energy revolution as one where they can be first and can do very, very well."[17] Chinese companies have adopted a classic mercantilist model—familiar from Japanese and

Korean electronics companies, among others—of using government-backed access to inexpensive and ready finance and a close relationship among business, universities, and government research centers to quickly increase manufacturing capacity. Profits have been less important than growth and sheer size—a strategy that is likely to pay off for the country's overall ambition to become a significant global player in renewable energy.

As recent developments in China show, public opinion is important in underpinning change, but politics matter more: it is political decisions that create the prices, taxes, and penalties that incentivize individual and—crucially—corporate behavior. Asian governments typically implement change through companies, whether private-sector, hybrid, or state-owned enterprises. In the chapters that follow, we will look at various instances of business-led innovation, which is a little-noticed but critical part of Asia's environmental transformation. Government and civil society play vital roles, but it is well-run businesses that have the manpower, the money, and the knowledge to reshape Asia's approach to the environment. Increasingly, Asian companies will be in the vanguard of change.

From Chinese solar panel manufacturers to Indian automakers to Indonesian palm-oil plantation owners, Asia's innovative and dynamic companies are using the challenge of sustainability to become even more competitive. They are often disruptors, their innovations changing the competitive landscape in their home countries and frequently upending global business as well.

In the chapters to come, we will look in depth at some of these innovative companies, with the aim of understanding the broader role of business in the shift to a greener Asia. Some of these companies already are known in the United States, such as carmaker BYD, 10 percent of which is owned by Warren Buffett's Berkshire Hathaway. Others are little known abroad, even though some have globally significant ambitions. Take CLP Holdings, one of Asia's largest private-sector utility companies; today, it has one of the world's most far-reaching plans to dramatically reduce the carbon intensity of what was, until a few years ago, a company made up mostly of coal-burning power plants.

As these two companies—and others profiled throughout the book—show, there are many paths to sustainability. Some companies, such as Indian car manufacturer Reva, were built with sustainability in mind from the beginning. Others are integrating sustainable practices into their existing business, quietly becoming environmental innovators through their internal operations. Hong Kong–based textile company Esquel, for instance, has emerged as a pioneer in smart water use in order to stay ahead of what management expects will be ever-tighter restrictions on water use in China. Indonesian palm plantation owner ANJ built a biomass electricity production plant fueled by waste from its palm processing operations, part of a wide-ranging commitment to show that palm plantations can be run in an environmentally sustainable fashion. Still others are expanding their core business to encompass new sustainable initiatives. Japanese finance company Orix has committed $1.1 billion to finance solar farms and rooftop installations. Singapore-based city builder Singbridge is constructing the energy-efficient Tianjin Eco-city in China.

Although the companies that we look at in the coming chapters operate in a wide range of industries and have a diverse set of strategies in their green pursuit, most of them are fairly large, reflecting the importance of large firms in Asia. Large companies, rather than fledgling ones, are most likely to succeed in growing a portfolio of environment-related businesses. Part of this behavior reflects the relatively rapid pace of economic growth in much of Asia, where new business areas can most easily be grasped—and dropped, if necessary—by larger companies, which enjoy better access to financing and have more managerial and technical talent. China is an exception because some of its largest renewable-energy companies began as start-ups; however, far-reaching government support in the form of money, land, and tax holidays, allows businesses that take off to grow extremely rapidly. Chinese companies that succeed tend to grow very large very fast.

~

The East is black, for we are still in the early years of this ragged, uneven process of change.

The shift to a green economy has been vastly overstated in the short run, but its long-run impact is probably just as grossly understated.[18] So the question then becomes this: What does the environmental future of Asia look like?

It is, of course, possible that Asia could suffer policy-making drift, a slide that would see ever-worse pollution and perhaps even water wars and a general breakdown of society. There are good reasons to be pessimistic, and many others have written and spoken eloquently about various parts of Asia's environmental challenge. Notable among them is Elizabeth Economy, an expert on China at the Council on Foreign Relations, whose *The River Runs Black* is a seminal book on modern China's environmental problems. More than a decade after her book was published, Economy remains skeptical about any rapid solutions to China's environmental problems. She notes that China's leaders have for a long time been bigger on talk than action; have prevented the development of the transparency, rule of law, and official accountability that are necessary for strong environmental protection; and have left standing flawed urbanization and industrialization policies that guarantee continued systemic environmental problems.[19]

Economy's skepticism to date has been well placed, but Asia's track record over the past fifty years makes it possible to put forth a green scenario that would see its environment improving dramatically over the next one to two decades. Asians, Asian governments, and Asian companies are driven to succeed, governments are under increasing pressure from citizens who may threaten their ability to govern, and there is broad agreement that their environmental problems can and must be fixed. The awareness of the problem is there—a recent survey shows that although the average person in China or India has a far smaller environmental impact than the world average, consumers from those two countries feel guiltier than any others about their impact on the environment.[20] In Asia's environmental crisis lies opportunity: its innovative companies are developing far-reaching responses to these environmental challenges. Spurred by its governments and supported by its cash-rich banks and capital markets, leading companies can use opportunities presented by their environmental nightmare to change the global economic landscape and to even overtake Western competitors in areas like solar and wind power, electric vehicles, and green infrastructure.

Countries like Japan and Singapore are setting the pace; substantial progress can be made quickly if Asian countries can muster the political will and set the proper market-oriented incentives to radically improve their physical environments. Asia's population and resource pressures leave its countries with little choice but to respond. Authoritarian or democratic, governments need to meet an ever-more-affluent citizenry's demands for cleaner water, cleaner air, and more livable cities. This transformation is already under way. If the experiences of Japan, Singapore, and Korea are any indication, much of Asia—including China—has the opportunity within a generation to make the shift to energy-efficient countries that embrace the idea of an environmentally friendly society as part of their national identities.

1

Energy: *Sun, Wind, and the End of Coal*

We may well call it black diamonds. Every basket is power and civilization. Coal is a portable climate. It carries the heat of the tropics to Labrador and the polar circle; and it is the means of transporting itself whithersoever it is wanted. Watt and Stephenson whispered in the ear of mankind their secret, that a half-ounce of coal will draw two tons a mile, and coal carries coal, by rail and by boat, to make Canada as warm as Calcutta, and with its comfort brings its industrial power.

—RALPH WALDO EMERSON

If this was a murder mystery, coal would be the villain hidden in plain sight.

A century ago coal powered the trains and ships that ushered in the modern age. Today, it still plays an outsized role in fueling the electricity plants on a continent whose people are literally coming out of the darkness—out of the enforced darkness of poverty, where streets are dim and students study by a kerosene light—into our electricity-driven age.

Coal is plentiful and cheap: coal-fired electricity powers eight of every ten light bulbs in China, a country that burns almost half the coal used worldwide every year.[1] Coal is an ever-ready servant, abundant and easy to use, and a steady and reliable producer of power.

Coal is also dangerous: it is the single largest cause of the air pollution that prematurely kills more than 1.2 million people each year in China alone, in addition to the more than 1,000 coal miners who perish in accidents in a typical year. Coal is also responsible for worsening climate change, accounting for more than 40 percent of greenhouse gas emissions.[2]

Coal is king in Asia. The continent accounts for almost two thirds of global coal use, up from a quarter in 1980. Coal use in Asia quintupled from 1980 until 2010, even while it fell in the rest of the world; still, coal burning in China is not expected to peak until around 2030. It is just too cheap, too easy, and too efficient at turning its latent caloric energy into the heat that drives the turbines that produce the electricity that Asia so badly needs; coal power is almost irresistible.[3]

The key to solving Asia's energy problems and its environmental nightmare lies in using less coal. China is making progress by adopting cleaner coal technologies. Indeed, a good part of China's attempts to reduce air pollution and slow the growth of carbon emissions will involve more efficient use of coal—building more efficient power stations and eliminating coal in smaller-scale industrial boilers.[4] But the more quickly China and other large, fast-growing countries like India and Indonesia can end their dependence on fossil fuels, especially coal, the faster air quality will improve and the easier it will be to mitigate the effects of climate change.

The end of the coal era will require a mix of solutions. One fuel that is the focus of a good deal of debate in China is natural gas, both conventional and unconventional (shale gas). Although natural gas is certain to play a larger role in meeting China's energy needs, it is unlikely to have the sort of transformative impact that it did in the United States. China has an undeveloped gas pipeline network; it spans only 4,500 kilometers, compared with 360,000 kilometers for the United States. Even China's 2015 target of 56 gigawatts (GW) of gas-fired generating capacity, up from 32 GW at the end of 2013, is substantially less than wind-powered generating capacity was at the end of 2013. China's energy and electricity prices are highly regulated, unlike in the United States, and China's gas prices are high relative to electricity prices. That means many gas-powered utilities must rely on financial subsidies from local governments. Although there is a lot of talk about the importance of developing China's shale gas reserves, these are unlikely to have more than a marginal impact on the country's energy picture. Even if China built out its natural gas pipeline network and increased electricity prices to make natural gas attractive to power plant operators, the need for vast amounts of water in unlocking shale resources makes it an unpalatable fuel for China. Moreover, the singular success of shale in the United States reflects a fragmented industry where small-scale, local wildcat operations have provided entrepreneurial drive. This industry structure is in direct contrast to China's extremely concentrated statist industry.[5]

Another important need is more efficient power grids. Here, as in many other areas, China is embarking on an aggressive expansion program. As noted in the introduction, it spent more in 2013 on emerging smart-grid technologies than the United States did. China now has a plan to build ten ultra-high-voltage power lines at a projected cost of $61 billion to bring power—including wind power—from its remote northern and western regions to its central and coastal cities.

Neither the increased use of conventional natural gas or unconventional shale gas nor the construction of a network of ultra-high-voltage power lines will change China's stark situation: it is a resource-poor, energy-hungry nation that will have to increasingly use energy more efficiently if its economic development is to continue. Even nuclear power, which China plans to grow from 14 to at least 58 GW by 2020, will make up only a small percentage of the country's electricity-generating capacity. That means China must continue to aggressively adopt renewable-energy sources such as solar and wind.

Solar and wind power, the focus of the next two chapters, are the two most dynamic and fast-changing sources of renewable energy. Although Japan is significant as both a manufacturer and a consumer of solar power and South Korea and Taiwan also manufacture solar panels, the dramatic changes that have roiled the global solar and wind industries in recent years are a China story. The Chinese wind and solar power industries have, since the turn of the century, transformed themselves from marginal players into significant global forces—in the case of solar, *the* dominant global force. China took advantage of its status as a technological latecomer to import foreign technology, often improving it, and drive costs down. It was able to do this because its land, labor, and capital are all underpriced for its favored industries.

Wind has an even more promising future in China. Wind power already accounts for a significant part of the country's installed electricity-generating capacity. With a strong political commitment and good policies, researchers at Harvard and Tsinghua believe that wind theoretically could account for all of China's electricity output by 2030 at a price comparable to that of coal.[6]

Still, despite impressive success in winning sales, both solar and wind show some of the problems of China's system. Although solar and wind are environmentally benign sources of power, as ongoing businesses many companies operating in this area are not sustainable. Competition among local governments, in concert with broad national policies, sparked extraordinary competition among these companies that led to a rapid decline in prices for solar and wind power. This price discounting was unintended—and completely at odds with China's notionally planned economy. It was wasteful, creating large losses that will be borne by ordinary Chinese, who as taxpayers ultimately pay most of the bill for the mistakes of their government and banks. This saga, in short, is not a textbook case of success but a tale of state planning—some of it successful and some not—coupled with the extraordinary, even reckless, ambitions of contemporary China's first generation of entrepreneurs. Unwittingly, however, the China model of a semiplanned industrial policy, buttressed by unnaturally low costs, has succeeded in making renewable energy cost-competitive far more quickly than even the most optimistic analysts would have imagined at the beginning of the 2000s, benefiting not only China but also the world.

1

The Sun Kings

> We are like tenant farmers chopping down the fence around our house for fuel when we should be using Nature's inexhaustible sources of energy—sun, wind and tide. I'd put my money on the sun and solar energy. What a source of power! I hope we don't have to wait until oil and coal run out before we tackle that.
>
> —THOMAS EDISON

In 1992, Shi Zhengrong completed his doctorate and found himself an expert in a field that wasn't quite ready for him. He'd studied physics at Australia's University of New South Wales, focusing on crystalline technology, the basic scientific building block of photovoltaic solar power. This knowledge, however, did not yet have much real-world application. Shi, originally from China, thought setting up a Chinese restaurant in Sydney was his best idea. As he told an audience in Hong Kong in 2008, his wife vetoed the restaurant idea and convinced him to look for work more closely related to his studies. He was able to stay in solar, working first at an academic post in Sydney—but real success followed after he started his own company. Shi returned to China and, with the help of local officials in the city of Wuxi, founded solar panel maker Suntech in 2001.[1]

Shi was successful beyond imagination. By 2005, Suntech had sales of $226 million. That year—just four years after it was founded—the company went public on the New York Stock Exchange (NYSE), raising $455 million in the exchange's biggest Chinese offering of the year. It also became the first private Chinese company to list on the Big Board. The next year, as a sign of the company's importance, Shi was invited to sit on the exchange's international advisory board.

Shi Zhengrong, founder and CEO of Suntech Power Co., at the World Economic Forum headquarters in Geneva, Switzerland, in August 2011. Photo credit: © AP/Martial Trezzini

For the next few years, Suntech's sales soared, and so did its stock market valuation. Encouraged by Suntech's success, other Chinese companies entered the industry, and the first decade of the 2000s saw Chinese companies sweep aside long-established international competitors. In 2012, PricewaterhouseCoopers released a report that ranked solar makers by their overall importance and gave Chinese companies eight of the ten top slots in the industry.[2]

Then came the fall. In March of 2013, after a series of missteps and a default on its bonds, Suntech's major operating unit filed for bankruptcy, bringing normal business operations at the parent to an end.

The story of how and why Suntech rose to such heights and then plummeted to earth holds lessons not only for solar panel makers and many of China's other green-tech industries but also for China's policy makers as they pursue their goal of increasing China's technological sophistication and advancing its hope of becoming one of the world's most innovative economies. Yet the story of Suntech's failure hides a more important tale, one in which China's impressive manufacturers have driven down prices far faster and further than anyone would have guessed a decade ago. Thanks to China's unique industrial structure, one that prizes growth over profitability, solar power is now broadly cost-competitive with other energy sources throughout much of the world. As solar panel prices plummeted, installations skyrocketed, with the global solar power base growing twenty-five-fold from 2005 through 2013. No company was more important to this transition than the tragically flawed Suntech.[3]

~

Suntech's growth was impressive. Within a decade of its founding, it went from an unknown start-up to the world's largest producer of photovoltaic solar modules—a corporate success that seemingly underscored China's newfound dominance of the clean-tech world.

Suntech's hometown of Wuxi is, by Chinese standards, a midsized city, though its population of more than six million people makes it larger than any U.S. city except New York. It is not far from Shanghai (a high-speed train now makes the journey in forty-five minutes) and lies in the Yangtze River Delta, a region that has proven to be fertile ground for China's economic growth. It is a particularly entrepreneurial part of the country, and officials there welcomed businesses and encouraged Suntech's wide-ranging ambitions.

For its first few years of business, Suntech's ambitions were consistently realized—and exceeded. Shi Zhengrong, rarely bashful, boasted in his 2006 letter to shareholders that the company had increased its manufacturing capacity twenty-seven-fold in the four years since it had started operations. The company shrewdly took advantage of an industry that was growing at a compound annual growth rate of 43 percent, with global solar sales mushrooming from $2 billion in 2000 to $9.8 billion in 2005. It was also outpacing opponents in keeping costs down. In the same shareholders' letter, Shi boasted that it would cost Suntech only $8 to $10 million to build a production line capable of producing 30 MW of solar panels a year because of cheap land, labor, and buildings. Some of Suntech's competitors, in higher-cost countries, would have to spend $30 to $75 million for similar capabilities.

The story was plausible. China's manufacturing companies have a well-deserved reputation for competent management and workers who are intensely, even obsessively, focused on bringing down manufacturing costs. Suntech, thanks in part to the cash raised from its NYSE stock sale, was able to invest and grow faster than its domestic competitors and remain much less expensive than its more established foreign peers. Aided by underpriced capital (a product of a state-run banking system that penalizes savers and pushes a glut of capital to investment) and inexpensive land (provided by local officials intent on bringing businesses to their area), the company had developed a recipe for exponential growth.[4]

Inexpensive products were key to Suntech's growth. The company had good technology, though it was in no sense on the leading edge of innovation. Indeed, China's success in solar came from using technology developed beginning in the 1950s at Bell Laboratories in the United States. Bell Labs, which functioned almost as a national laboratory even though it was owned by telephone monopoly AT&T, was at the forefront of the development of semiconductors, and solar power was an offshoot of that effort. Bell Labs's initial breakthroughs were buttressed by substantial support from the U.S. government, which spent heavily on the development of solar power.

The first practical use of solar power was for satellites. Solar allowed for a longer useful life than batteries alone did, and with cost of secondary importance in the early decades of the space race, scientists were able to use the still-expensive technology. The Vanguard I satellite, covered in solar cells, was launched in 1958. In the late 1960s, scientist Elliot Berman discovered that scrap silicon from semiconductor manufacturing, even with imperfections, could be used for solar power. Exxon's backing of Berman's discovery lowered manufacturing costs, and solar garnered significant interest as rising oil prices after the 1973 OPEC oil embargo sparked concerns about energy security. A series of innovations since then has continued to push down prices. The price per watt of electricity produced by solar has fallen almost 99 percent since the 1970s, from $70 a watt to 80 cents in 2012.[5] By the time Suntech entered the scene, the technology was far enough along that it was able to grow successfully without needing to be more than incrementally innovative.

The dramatic increase in solar power just after the turn of the century owed much to government subsidies, such as those in Germany, Spain, and Italy, that provided incentives for companies to build solar farms by guaranteeing a high price for their power output. With the solar market growing rapidly, thanks to these subsidies for solar installations, the markedly lower prices charged by Suntech and other Chinese companies allowed them to quickly take market share from their high-cost Japanese, German, and American competitors.[6]

From its early days, Suntech—like so many Chinese companies—focused on exports, counting on its low costs to win sales. With the global solar market growing rapidly, especially where guaranteed high selling costs for solar power stimulated the market, the company was in an enviable position. In 2007, it opened its U.S. headquarters in San Francisco. The next year saw sales offices open in Germany, Spain, South Korea, and Australia. It built a global network of projects at sites ranging from the San Francisco Airport to the Bird's Nest Stadium at the Beijing Olympics to the Yas Formula One race track in Abu Dhabi. Its solar panels went up on a school in Lebanon and a Harrah's casino in southern California.

In the two years after its 2005 initial public offering, Suntech's shares increased in value almost sixfold, flying high on the company's ability to be a low-cost manufacturer in a fast-growing industry. The company's market value on the NYSE peaked at $13.3 billion in December 2007.[7] Shi soon was celebrated as the face of an entrepreneurial, more environmentally conscious China, one whose companies were not just churning out iPods and Barbie dolls but also taking their place among global clean-tech leaders. In 2007, Shi was named one of *Time*'s "Heroes of the Environment" and China's Green Person of the Year. He also became one of China's richest men, with *Forbes* pegging his net worth at $2.9 billion in 2008. That same year the Asia Society named Shi as one of its three "climate heroes," along with Alcoa Chairman Alain Belda and R. K. Pachauri, chairman of the Nobel Prize–winning Intergovernmental Panel on Climate Change. In 2009, the World Bank's private-sector unit, the International Finance Corp. (IFC), gave Suntech its seal of approval by loaning it $50 million in the form of a bond convertible into company shares. The Massachusetts Institute of Technology's *Technology Review* included Suntech on its 2012 list of the world's 50 most disruptive technology companies, citing its success in manufacturing low-cost solar cells. By 2011, six years after the company went public, its sales had soared to $3.1 billion.[8]

Suntech benefited from significant support from government, especially local governments, and from banks, including the massive, policy-oriented China Development Bank. However, its success was not simply a story of inexpensive Chinese capital. The company was also able to tap international sources of funding, such as the NYSE and the IFC. That a start-up company in China could raise almost $1.9 billion in five years from global investors reflects the dramatic globalization of finance in recent decades.[9] Suntech used these low-cost funds to drive improvements in photovoltaic efficiency as well as to reduce the total production costs of solar modules.

However, Suntech's rapid growth—paralleled among other hopeful entrants into the solar market—also sowed the seeds of its fall. In spurring the expansion of the domestic solar industry, it helped create an environment of chronic overinvestment, with an unsustainable number of companies chasing sales growth with little regard for profitability. Suntech and China's solar industry as a whole have shown that they can disrupt markets, and they have done the world a service by accelerating the introduction of renewable-energy technologies. But the corporate success of these companies has been ephemeral. Many have been unable to build sustainable businesses.

Suntech thrived when polysilicon (formally known as polycrystalline silicon), a key ingredient in solar panel manufacturing, was in extremely short supply. As one of the first Chinese entrants into solar and driven by Shi's technological know-how, the company had signed long-term supply contracts guaranteeing it access to polysilicon. It did this in the early 2000s, when the solar industry was much smaller. With the industry's rapid expansion, especially as other Chinese solar makers entered the business, the small number of polysilicon makers simply could not keep up with demand. Suntech had guaranteed its polysilicon supply, leaving its rivals struggling to buy the leftover scrap material from spot-market traders, often at far higher prices than Suntech paid.

Predictably businesses responded to the shortage and the high prices by building new polysilicon manufacturing plants. Korea and China started building their first dedicated polysilicon factories in 2006, with significant production dramatically driving down prices in 2009. From an overall economic perspective, investing in polysilicon manufacturing capacity made little sense for China. Polysilicon production is energy-intensive, and China's electrical grid was already struggling to keep up with the country's rapid growth. It is also a sophisticated process that had previously been limited to manufacturers in Europe, the United States, and Japan. By investing in polysilicon manufacturing, China was importing yet another energy-intensive industry from the West—one fueled by inexpensive and highly polluting coal-fired electricity.

Yet from the standpoint of individual companies struggling to buy polysilicon, manufacturing the material seemed sensible. "China had no alternative," says Charles Bai, chief financial officer of up-and-coming ReneSola in the early and mid-2000s. ReneSola was one of the many Chinese solar companies that felt they had no choice but to invest in the business because of the difficulty they had in buying polysilicon, no matter what price they were willing to pay. Major international firms "simply wouldn't sell to us" because all of their production was promised to other companies under long-term contracts. The leftovers they bought were often of poor quality. ReneSola employed hundreds of people to clean up scrap silicon, much of it broken wafers, that Samsung, LG, and other semiconductor companies sold to ReneSola for almost $500 a kilogram at the peak in 2008. Moreover, "I heard enough about how difficult polysilicon [manufacturing] was and how Chinese manufacturers would never catch up to think people were trying to keep China from getting polysilicon," says Bai.

Bai, a Chinese-born, Swiss-educated Canadian citizen, spent two months during the summer of 2006 in London, where he raised $50 million on the London Stock Exchange's AIM, a NASDAQ-like bourse for up-and-coming companies. "Fifty million dollars didn't go very far," remembers Bai. "We needed $50 million for the furnaces we had ordered. Then we expanded and needed more furnaces." This fits into a larger pattern, where the imperative to expand and the ability to raise substantial amounts of capital drive China's investments.[10]

Manufacturing polysilicon is like making a semiconductor and involves various complex techniques—many proprietary—but Chinese solar companies mastered them in short order, thanks to their access to money, their base of good scientific and engineering talent, and the industry experts from the West who served as consultants. Chinese companies showed that they were adept at absorbing—and even improving—manufacturing processes invented in more-developed countries.

These new polysilicon plants were a source of pride to companies and local officials. Calvin Lau, now the chief financial officer at Ming Yang Wind Power Group but previously in the solar industry, remembers the excitement at the inauguration of a polysilicon plant. At one such event, he recalls a company executive showing a newly produced polysilicon ingot and saying, "That's worth $1 million"—to the delight and astonishment of local officials.[11] These sorts of visible successes meant that money flowed into polysilicon production. Local officials could help companies obtain inexpensive land in China's ubiquitous industrial parks, and they could encourage banks to lend money to build factories.

As a result of polysilicon fever, prices plummeted 95 percent in five years, from $350 or more a kilogram in 2008 to around $17 a kilogram in 2013.[12] This astonishing drop in price transformed Suntech's long-term contracts from a strength to a crippling weakness. Most of its contracts were "take-or-pay" agreements, obligating the company to pay for the production at a predetermined price, whether it took the material or not. Suntech had factored a more gradual price decline into the contracts, so it found itself with more and more polysilicon that it had to buy at above-market prices. In 2011, it terminated a ten-year take-or-pay polysilicon contract early at a substantial cost to the solar panel maker—$120 million.[13]

This glut of polysilicon manufacturing capacity not only rendered Suntech's contracts disadvantageous but also led to a global decline in the price of solar panels. These dropping prices revealed another of the company's weaknesses—its debt burden. It had been extremely nimble in its ability to raise money in the international financial markets after its 2005 NYSE listing. In 2007, it sold $500 million of convertible bonds with an interest rate of just 25 basis points (0.25 percent). These bonds were, as their name implies, convertible into company stock; investors were willing to accept a low interest rate in return for the chance to buy shares at a predetermined level if the company did well. Other stock and bond offerings followed in 2008 and 2009 (including the offering bought by the IFC), raising a total $912 million.

Although Suntech survived the 2008–2009 global financial crisis, thanks to its ability to keep raising money from banks and the stock and bond markets, its size had made it extremely vulnerable to faster-than-expected declines in solar panel prices. As the average selling price per watt of solar photovoltaic modules fell from $3.72 in 2007 to $1.51 in 2011,[14] the company's debt burden put it in a position similar to that of a wealthy homeowner in the midst of a massive expansion of his vacation house who suddenly finds himself with no annual bonus, a salary cut, and no savings. Between 2005 and 2011, Suntech's sales had ballooned from $226 million to $3.1 billion—but its long- and short-term debt had also risen, from $56 million to $2.3 billion. Its interest expense had grown from a mere $8 million in 2005 to $136 million in 2011.

It became clear that Suntech had grown too fast. The company had increased its annual manufacturing capacity of solar cells exponentially over the course of the decade, from 10 MW in 2002 to 1,800 MW by 2010. This required large-scale investments in new wafer production plants, and from 2008 through 2010, Suntech spent $815 million on new factories.[15] Because it was growing so quickly, the company needed cash to finance its growing inventory of raw materials and finished products, and it also needed cash to tide it over until its customers paid it for goods. As long as sales kept increasing rapidly and the company could keep raising money by selling shares, raising debt, and borrowing from banks, its rapid expansion could continue. But as sales began to level off, it suddenly found itself defenseless.

Making matters worse, Germany and Spain—Suntech's two largest markets and countries whose solar-power subsidies had provided much of the fuel to ignite the demand for solar panels—cut the prices they paid for solar power in the face of increased fiscal pressures and the falling prices of new solar panels. Sales to these two key markets fell in 2011, bringing the company's overall sales growth almost to a halt. Sales had nearly doubled in 2010; in 2011, they were up an anemic 5 percent.[16]

By 2011, Suntech was struggling to make the interest and principal payments on its debt. The cash generated from its business operations was five times as much as its interest expense in 2005, but by 2011, there was not enough cash from operations to pay the interest on its debt, let alone to start paying down the principal. In 2011, collapsing solar panel prices left the company staggering under a $1.0 billion loss. In the first three months of 2012, the last time it reported comprehensive financial results, it lost another $133 million.[17]

Suntech was scrambling for survival. For all of its impressive top-line sales growth, the company was imprisoned in a cycle of profitless prosperity. The relentless demand for money to build new manufacturing plants diverted so much of its revenue that it couldn't keep paying the interest on its debt or even generate the cash to survive as an ongoing business. Warren Buffett is fond of saying that when the tide goes out, we can see who is swimming naked. The tide went out, and despite never having paid investors even a penny of its profits in the form of dividends, Suntech was financially naked.

In March of 2013, Suntech defaulted on a $541 million convertible note, triggering defaults on its remaining debt. Five days later, its principal operating subsidiary in Wuxi filed for bankruptcy, throwing the company into financial and managerial turmoil. When Jiangsu Province formally accepted the petition for insolvency and restructuring of the Wuxi Suntech Power subsidiary, a group of creditors, mostly Chinese banks, was owed as much as $1.14 billion. The banks included two of China's largest, ICBC and the Agricultural Bank of China, as well as the IFC. Adding to Suntech's woes was the discovery that a €560 million bond, supposedly posted as collateral guaranteeing payments to China Development Bank on behalf of a European joint venture partner, did not exist. Shi Zhengrong was forced out as chairman and CEO, though he remained on the board of directors. The company's market valuation fell below $70 million in April 2013, down more than 99 percent from its peak of $13.29 billion. An investor who unluckily bought $100,000 of Suntech's shares at their high at the end of 2007 would have received just a little more than $500 if the shares were sold at the spring 2013 trough.[18]

After the bankruptcy of its Wuxi Suntech unit, conflict arose among the members of the board of directors as the company attempted to reorganize. In August 2013, three of the company's six independent directors resigned. Typically, unhappy directors resign quietly, issuing anodyne statements designed simply to meet regulatory requirements—but not at Suntech, where the three departing directors issued a brutally frank and wide-sweeping departure note. The note cited their reasons for leaving: "Severe cash flow drain with unclear prospect of securing new capital; difficult prospects on completing consensual restructuring with convertible bondholders; lack of clear business plan; loss of critical talent and potential severe HR [Human Resources] retention issues; failure to pay outside legal counsel; potential erosion of internal controls; and impairment of employees' ability to function effectively." In late 2013, the company announced plans to bring in a new strategic investor.

Many questions remain about Suntech, questions that may be resolved as a result of lawsuits filed against the company and against Shi Zhengrong, a man who has gone from being lauded as the Sun King to being denigrated as a free-spending megalomaniac who was led astray by his initial success. But to paint Shi as a villain or as uniquely greedy or to seize on allegations of irregularities—as many in the industry do—is to obscure the larger issues. Despite China's limits on deposit interest rates, the country's financial system generates excess savings because savers have few alternatives for their money. These savings in turn need to be recycled in the form of loans. High investments, often with little capital discipline, are the result.

The Austrian economist Joseph Schumpeter pioneered the concept of creative destruction—of new companies innovating and laying waste tradition-bound incumbents. Suntech, which had so successfully disrupted the global solar industry by building so much capacity, found itself undercut by that very same situation in the polysilicon area—excess production capacity. The company was lauded for its ability to disrupt, but it was more destroyer than creator, a tornado that ravaged many of its competitors before exhausting itself. In a country that is distorted by too much capital and not enough productive investment outlets, the plunge in polysilicon prices reflected Chinese-style market forces at work. This counterproductive behavior was characterized by identification of and investment in an industry with high profit margins and short supply; significant but not insurmountable barriers to entry in the form of capital and technology requirements; success by an initial leader or group of leaders; and a wave of undisciplined investment by me-too companies, aided by local governments and cash-rich lenders. In the end, companies were unable to successfully differentiate their products except by cutting prices.

Suntech shows the strengths and weaknesses not just of China's solar industry but also of the country's approach to technology acquisition and manufacturing. Shi Zhengrong understood that Chinese manufacturers had low enough costs and good enough technology that they could undercut established solar companies. Suntech and many of its industry peers grew from literally nothing to become among the largest companies in a fast-growing industry because they were able to manufacture a technologically mature product at a price far below those of their European, American, and Japanese rivals. This combination of good technical, engineering, and manufacturing skills; ability to grasp a market opportunity; and access to a large amount of financing is what makes Chinese companies so important in shaping the future of much of the global renewable-energy industry.

Yet this model in some sense carries within it the seeds of its own destruction. The access to easy financing means relatively little discipline in investments. This leads to overinvestment and a tendency not to worry too much about profits. When the business cycle turns against these capital-intensive companies, they are particularly vulnerable, as the Suntech story demonstrates.

Unsustainable though Suntech was as a business, the company's real-world achievements should not be ignored. In 2011, it became the first solar panel maker to have installed a cumulative total of 5 GW of solar capacity. By 2012, it had installed 7 GW of solar power, almost equal to the total electricity-generating capacity of Ireland.

Shi Zhengrong, whatever his flaws, was a visionary pioneer who took advantage of a Chinese economy that repeatedly produces investment bubbles, in renewable energy and other areas. This is not good news for investors or for Chinese taxpayers, who have to pay the bills for the inevitable bad loans that result. But it has ensured that wind, solar, and other renewable forms of energy have been adopted more quickly than would otherwise have been the case. Suntech has not proved to be a sustainable business. But Shi Zhengrong and his company, in taking advantage of China's structural need to invest with little regard for profitability, dramatically accelerated the spread of solar technology around the world and permanently changed the economics of the renewable-energy industry. The challenge for China then becomes finding a more sustainable economic structure that will allow it to continue promoting this kind of growth in new industries without also ensuring the demise of the businesses that operate within them.

~

The problems at Suntech show in microcosm everything that is so impressive about China's economic growth as well as why China could ultimately prove incapable of moving to the next stage of economic development.

One can think of China's industrial structure as "start-up statism." China fuses some of the world's most risk-craving, dynamic entrepreneurs with top-down state policies with the help of government-directed capital. Suntech exemplifies what can happen to a company when faced with these forces. Japan and South Korea pursued similar patterns of what economists call financial repression (a policy that lets government issue debt at low rates by denying savers high returns) coupled with government-guided, investment-led, high-growth models, but neither country allowed the sort of anything-goes frenzy seen in China's solar industry, where hundreds of competing companies started operations within the space of a few years. Indeed, China has several notable characteristics that separate it from Japan and South Korea in their high-investment, high-growth years.

First is the sheer amount of capital available. Bloomberg's New Energy Finance Unit calculates that Chinese financial institutions had committed $47.5 billion in loans and lines of credit to the solar industry alone as of September 2013. It estimates further that the ten largest Chinese solar companies had $28.8 billion in debt, most of it owed to Chinese banks and other institutions.[19] This leads companies to pursue a "bicycle" strategy of growth—keep pedaling faster, or risk falling over—which has been common in developing Asia. Many companies have used access to inexpensive capital, usually bank loans, to try to grow big enough, fast enough, to develop a lasting business. Korea's big business groups, the *chaebol*, followed this strategy with great success—until the 1997 financial crisis brought down several of them, most prominently, Daewoo, once the country's fourth largest business empire. But China differentiates itself from its neighbors in terms of the sheer amount of capital available.

China's economic system is designed to punish savings and reward investment, which is partly the reason for this glut of capital. Under these conditions of financial repression, Chinese savers are forced to subsidize investors because, like Japanese and Korean savers before them, they have few alternatives for their money. Restricted from moving money abroad by capital controls, they can put it in the bank, where deposits typically earn less interest than inflation. They can put it in the stock market, although long-term returns (adjusted for inflation) have been minimal, notwithstanding the country's dramatic economic growth. This has forced Chinese investment into other channels—notably, the property market, where the flood of liquidity has led to rapidly increasing property prices in China's main cities. In fact, such rapidly increasing property prices, coupled with large numbers of empty apartments, are a typical outcome of financial repression. Hard assets such as real estate, and sometimes gold, tend to become favored investments and a trusted store of value, rather than things to be lived in and used.

Because the economic system has few outlets for domestic savings other than bank deposits, stocks, and property purchases, much more money flows to bank deposits than would otherwise be the case. Banks, for their part, must use the money that depositors have entrusted to them. They would lose money if their deposits were left idle, so they recycle these deposits in the form of lending. Consumer lending is discouraged by government diktat, so consumer credit is virtually nonexistent, except for mortgages. (In China, consumer spending accounts for roughly one-third of annual economic growth; in the United States, consumers are responsible for about two-thirds of economic growth.) Instead, most of the money lent by banks is invested in capital projects—in roads and highways and in facilities like steel mills, cement factories, and solar plants.

As a result, China's savings rate and investment rate, two sides of the same coin, are both extraordinarily high, leading to excess funds available for investment and, ultimately, to a large degree of inefficient, wasteful investment.[20] Although Japan and South Korea pursued similar unbalanced growth models, penalizing savers and maintaining high levels of investment, no major country has ever saved or invested as much as China. Investment has accounted for at least 30 percent of China's annual economic growth since the early 1980s, when Deng Xiaoping's economic reforms began in earnest. Since the turn of the century, the figure has generally been above 40 percent—extremely high by any historical standards. In 2012, the figure was 48 percent. No significant economy has ever witnessed anything like this investment level outside of a wartime period.

This extraordinary capital investment is leading to fears of a credit-fueled bubble, as new loans have grown far faster than overall economic growth over the past decade.

The second characteristic that makes China unique among Asian countries in the region is the structure of its local governments, which enables them to help fast-growing companies like Suntech. Local officials' promotion prospects are to a large degree driven by the economic growth and the jobs created in their areas. At the same time, local officials have a good deal of discretion over land and taxation. They can easily convert land to industrial use (often seizing from existing users who are paid derisory compensation) and use it to attract businesses, frequently adding a variety of incentives in the form of reduced taxes and utility and water charges.[21] They can also informally encourage bank lending and other forms of financial support. In 2005, the year of its NYSE listing, Suntech paid an effective tax rate of 10.9 percent of its profits. In the next six years, its tax rate averaged 5.8 percent, excluding two years in which it did not pay any taxes at all. Meanwhile, China's top corporate tax rate was 33 percent until 2006, when it was cut to 25 percent. The company never paid even one-third of the standard corporate tax rate.[22]

Finally, there is the me-too quality of Chinese companies. A particular strain of entrepreneurship has developed as a result of the Chinese system. Successful businessmen are guided by an overriding drive to grow and thus have the scale to better deal with regulators and banks, rather than focusing on technical innovation and other, more mundane operating aspects of well-run businesses. China's high economic growth has rewarded this sort of boldness, and in the past three decades, China has produced some of the world's most audacious entrepreneurs. The 2014 Hurun Global Rich List counts 358 Chinese billionaires, second only to the United States.[23] These fortunes largely are built on physical assets, especially property and manufacturing. Chinese businessmen who can obtain cheap loans and access to inexpensive land are eager to jump into areas that are proven winners, rather than trying to pioneer new ones. The combination of a need to invest savings and the powerful incentives for local officials to encourage business growth leads to quick and widespread imitation of successful firms like Suntech. Yet, particularly given the continuing weakness of intellectual property rights in China, it also virtually guarantees that long-term profitability will be destroyed for all but a handful of survivors. The brutal reality is that, within China, larger macroeconomic and political factors force companies to invest heavily without winning any sustainable competitive advantage. Success almost inevitably is copied, so profitability is just as surely destroyed.

By 2013, in one of those figures that defy belief, half of China's 600 cities had at least one solar producer, according to the China Renewable Energy Society. One industry listing put the number of Chinese solar manufacturers at 519 as of mid-2014, about as many as in Europe, the United States, Japan, Korea, and India combined.[24]

Profitability suffers because of these high levels of copycat investments. From 2005 to 2010, operating profits of the ten largest photovoltaic manufacturers ranged between about 15 percent and 17 percent, with a dip to 7.9 percent in 2009, the year of the global financial crisis.[25] After 2010, profit margins were obliterated. According to an analysis by PricewaterhouseCoopers, the ten largest solar companies reported an operating loss in 2011 equivalent to 1.2 percent of sales. The actual loss was higher because operating losses do not include interest expenses or taxes.

If they run into trouble, favored companies are typically rescued with more bank loans or other forms of financial support—so the natural impulse is to try, as Suntech did, to become too big and too important to fail. This is an extreme form of moral hazard, in which extreme risk-taking—gambling the company—is implicitly encouraged by a broader system that bails out large failures. This failure of market discipline makes China's economic growth inherently more volatile.

The solar industry showed the limits of China's central government. Although China is a good example of what development literature calls a strong state, where government can set and enforce rules with relative autonomy, the country is simply too big and there are too many conflicting interests even within government itself for national economic policies to consistently prevail. In industries that are particularly capital-intensive and in which the state feels it has a strong national security interest, national policies usually stay in force; telecommunications, utilities, and petroleum are all examples of this. But even here, there are limits to Beijing's power. Local interests often determine whose electrical solar or wind power gets dispatched, and thus who gets paid, from among myriad power producers. This ground-level favoritism has had significant negative consequences for some solar operators.

Similar examples are apparent in numerous Chinese industries. Automobiles are an instructive case in demonstrating the limits of central government control. For more than a decade, Chinese central planners have been trying to winnow the number of auto manufacturers, which still number over one hundred. Yet local policies have kept many marginal automakers in business. Those polices include protectionism (mandating the purchase of locally made cars for government departments, local government-owned or -linked businesses, and taxi fleets) and more subtle forms of favoritism, such as tax waivers, reductions in electricity and other utility charges, and cheap land.

Steel is another case in point. Yu Yongding, a Chinese academic with close ties to the government, notes that the profit from two tons of steel is about enough to buy a lollipop. That situation exists thanks to overcapacity driven by China's policy of underpricing capital, land, and labor.[26] Cement has significant overcapacity prompted by the same forces—too much capital in a situation of financial repression and the discretion allowed local officials. Industry after industry, from glassmaking to shipbuilding, is beset by similar problems.

In short, solar power exemplifies the impact that the distorted characteristics of China's economic growth have on individual industries and the ways in which activities that are rational at the local level can lead to unsustainable excesses at the national level. All economies, and especially fast-growing economies, are prone to cycles. The Chinese economic system as it is currently structured is particularly prone to extraordinary bouts of wasteful investment. Although these destroy profitability, they often, and certainly in the case of solar, bring prices down and lead to a diffusion of new technologies far faster than would be the case with a more orthodox approach. In an economy whose defining characteristic for decades, even centuries, has been scarcity, encouraging the animal spirits of the nation's hungry businessmen with inexpensive loans and a variety of supportive policies almost inevitably leads to overinvestment. Firms desperate to book sales in a struggle to survive inevitably slash prices.

Given all this, it seems that the rapid expansion of solar was not an accident but an imperative. It was all but inevitable in light of the financial repression driving banks to invest, local officials' desire for economic growth, national and international policies designed to promote solar power, and the sheer size of the country. What's more, it's likely that this cycle will be repeated, in other industries, for some time to come. In the case of solar panels, the consequences for consumers worldwide are positive, but the challenges for global corporate competitors have often been ruinous and have led to bitter trade disputes between China and the United States. (Ironically, punitive tariffs and duties imposed by the United States and the European Union in 2013 stabilized prices and helped avert further financial damage for the Chinese industry.) In the long term, China's enthusiastic investment in different sectors could lead to both rapidly dropping prices and a stunting of technological innovation as older, easy-to-master technology is used for longer than it would be otherwise.

This is a systemic challenge. Officials in Beijing want Chinese solar firms to succeed, and they would like an orderly industry, not one prone to destructive price wars. Yet Beijing is largely unable to stop this roller coaster of capacity surges and pricing falls.

It will be instructive to watch China's solar industry in the aftermath of Suntech's collapse. Wuxi Suntech was bought out of bankruptcy—for ¥ 3 billion ($465 million) in cash—by a low-profile Hong Kong–based investor Zheng Jianming (in Cantonese his name is Cheng King-ming). Zheng got his start as a researcher for China's State Council and then moved into real estate investing in Shanghai and Hong Kong. His Hong Kong–listed company, Shunfeng Photovoltaic, has undergone a transformation as Zheng positions himself to become potentially one of the most important players in the global solar industry. Shunfeng's 2013 annual report and subsequent press releases noted bank loans and other financing facilities totaling ¥ 21.2 billion ($3.6 billion) from the China Development Bank, China Mingsheng Bank, and China Merchants Bank. In 2014, Shunfeng also bought a 21.6 percent stake in another large, troubled solar manufacturer, LDK Solar; LDK declared bankruptcy later that year. As of mid-August 2014 J, Shunfeng had a stock market valuation of $3.1 billion, ranking it the eighth most valuable among the 128 solar companies tracked by a broad Bloomberg index.[27]

~

China has spent almost $50 billion in little more than a decade building the world's largest solar manufacturing industry. By some accounts, it has little to show for the effort. Its companies have yet to demonstrate that they are viable, at least as measured by their ability to earn profits. In fact, cynics could argue that China's solar effort is a mirage, built on subsidized electricity, most of it produced by coal-fired power plants and often using dirty manufacturing processes that have themselves resulted in environmental damage; financed with large amounts of inexpensive loans; and benefiting from preferential land and tax policies. In fact, that is similar to what was claimed in an antidumping trade suit filed against Chinese producers that resulted in a variety of punitive tariffs and other penalties imposed on the Chinese by the United States and Europe in 2012 and 2013.

But the extreme pessimism of this view is mistaken. China has succeeded in dramatically accelerating the spread of solar power globally. Its companies' willingness to sell at rock-bottom prices forced the industry to innovate and further drove down prices. Charles Yonts, an analyst at CLSA Securities in Hong Kong, calculates that solar prices fall 20 percent every time the total amount of installed solar power doubles.[28] China's low prices accelerated these price declines and also accelerated solar's entrance into the energy mix as it became broadly competitive with most other forms of conventional power.

Solar's cost is only about 1 percent of what it was in the 1970s, with most of the drop occurring in the past decade as a result of Chinese competition. As solar achieves the holy grail of so-called grid parity—meaning that users pay the same price for solar as they would for competing power—its popularity will grow. The notion of grid parity is enormously complicated to calculate in practice.[29] Broadly speaking, however, solar will be competitive with many forms of fossil fuels in half of the world by 2015.

Lower photovoltaic panel costs are only part of the solution to the problem of how to implement solar power on a widespread basis. In the United States, for instance, so-called soft costs—the costs of required permits and inspections, the labor needed for installation of solar panels, and financing—account for more than half the installation price for a residential system. Even if panels were given away for free, the average residential solar system would still cost about $20,000, estimates energy research think tank Rocky Mountain Institute. One study benchmarking these costs found that Germany's residential solar costs were half those of the United States. This difference, according to research conducted by the institute and the U.S. Energy Department's National Renewable Energy Laboratory, principally reflects these countries' differing levels of bureaucratic involvement in signing up new customers, getting permits, performing inspections, and creating interconnections. Germany, for example, requires only a single registration form for solar operators to qualify for utilities' subsidized solar purchases and only a single approval to connect to the electrical grid.[30]

Streamlining bureaucratic processes seems like one important step for stimulating the widespread adoption of solar power. Another is providing temporary subsidies. Europe's subsidies have made it by far the world's most significant location for solar power, with about 70 GW of installed solar capacity as of the end of 2012, about 70 percent of the global total at the time.[31] But Europe's strained fiscal circumstances, coupled with the unexpectedly high costs of the subsidies, have forced the curtailment of subsidies in some of the largest markets, such as Germany and Spain.

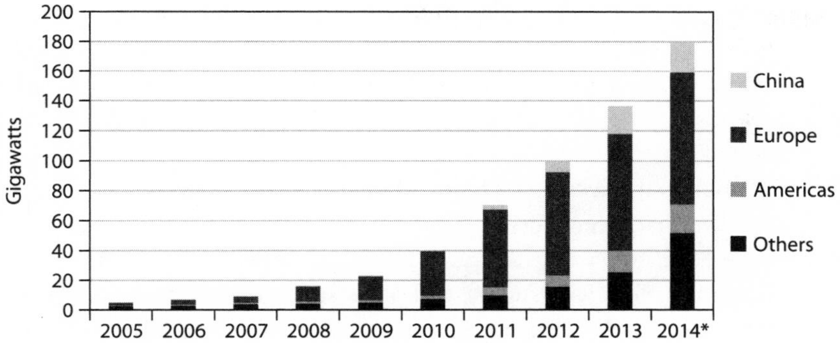

Global Solar Generating Capacity
New solar installations in China are now driving the market. (*) 2014 is estimated.
Source: EPIA, Mercom Capital Group

As European subsidies wane, China and Japan are supplanting it as major new sources of demand. In July 2013, the Chinese government set a goal to have 35 GW of solar-power capacity installed by the end of 2015—a fourfold increase from the total of 8.3 GW that had been installed by the end of 2012, a target that appears well within reach.[32] This will mean installing about 10 GW of solar power a year. For the sake of comparison, Switzerland's entire installed base of power is only about 18 GW. Solar companies such as Canadian Solar are increasingly moving beyond just manufacturing to set up their own solar farms. In the wake of the 2011 Fukushima disaster, Japan has found itself similarly bullish on solar power.

Japan has long toyed with solar power and has gotten more serious about it in the wake of the devastating "triple disaster"—beginning with the March 11, 2011 Great Tohoku Earthquake, one of the most powerful earthquakes ever recorded, and the resulting devastating tsunami with 133-foot waves, which together killed almost 16,000 people. The earthquake pushed the main island of Japan eight feet to the east and shifted the earth's axis between four and ten inches. The third part of the disaster came when emergency cooling systems at the Daiichi nuclear power plant failed as a result of the tsunami, causing the worst nuclear accident since the 1986 Chernobyl meltdown. The failure of the generators allowed a dangerous buildup of heat, leading to a series of hydrogen explosions and a meltdown of three of the Fukushima nuclear reactors. The nuclear disaster continued into late 2014 with continued leaks of radioactive water. The nuclear fallout was especially bitter for Japan, given its suffering after the atomic bombings of Hiroshima and Nagasaki.

Because Japan lacks fossil-fuel resources such as coal or natural gas, nuclear power had been a critical component of the country's energy mix. Before the Fukushima disaster, nuclear power made up about 26 percent of the country's electricity supply, and plans called for increasing nuclear's share to 45 percent. The nuclear accident has been a serious blow to Japan's self-image of technological infallibility. Searing reports in the aftermath of the crisis pointed to a culture of complacency as well as to a closed "nuclear village" that included utilities such as Fukushima's operator, Tokyo Electric Power Corp.; government regulators and bureaucrats; nuclear equipment manufacturers; legislators; the financial sector; and the news media.[33] These composed, in the words of scholar Jeff Kingston, "a village without boundaries or residence cards, an imagined collective bound by solidarity over promoting nuclear energy."[34]

On the eve of the 2011 Fukushima meltdown, Japan had fifty-four nuclear reactors producing about one-quarter of the country's electricity. Before the crisis, there had been hope that Japanese industry would benefit from an expected global nuclear renaissance. Now, Japan's nuclear industry is hoping simply to survive. In 2012, in the immediate aftermath of the crisis, the government of Prime Minister Naoto Kan pledged to phase out nuclear power by 2040. In 2014, with the long-dominant Liberal Democratic Party back in power, the government backed away from the pledge, but the continued difficulties at the Fukushima plant have left the Japanese public wary. All of the country's fifty-four reactors were shut down after the Fukushima accident, and only two reactors have operated even intermittently since then.

The loss of nuclear power has forced Japan to import fossil fuels, creating trade deficits and worries about the long-term impact of significant energy imports on the country's economy. Plans are under way to draw up a new energy policy for the country, one in which renewable sources will play a much larger role; the government hopes to boost renewables' share of the total from 10 to 20 percent.[35] Most of this is hydropower, but other renewable technologies are growing in Japan. Geothermal is an area that has potential, though it has not been developed to any significant degree. Wind is unlikely to develop a significant presence in Japan because of the country's shortage of large land parcels in places with consistent wind.

This leaves solar as the biggest beneficiary of this new energy policy. Japan remains one of the largest solar producers, with about 7 GW of the world's total installed base of 100 GW of solar generating capacity as of 2012, though its solar installations generated less than 1 percent of the country's electricity. In the fiscal year ending in March 2014, Japan installed 7 GW of solar capacity. Pre-Fukushima targets, set in 2009, called for installed solar capacity of a modest 28 GW by 2020 and 53 GW by 2030, but those targets look likely to be easily exceeded.[36]

Growth seems to be accelerating. In 2010, Japan had only 1 GW of installed solar capacity, while by mid-2014 it had over 14 GW. However, growth may slow if electric utilities can't accommodate this new solar power. The local utility on the northern island of Hokkaido says it will accept only 400 MW of solar power, meaning that many applications from developers, who had hoped to install an additional 1 GW of solar power on top of that limit, are likely to be refused. This parallels the pattern of many utilities worldwide that are reluctant to accept solar. Solar is fickle, so it cannot be relied on to produce power when it is needed; when solar power *is* produced, a number of technical measures must be taken to ensure that it is smoothly fed into the grid, as with any source of power. The nightmare for utilities is that solar power may be too successful, allowing it to displace utilities' own power plants. This is starting to happen in Germany and parts of the southwestern United States, notably Arizona, leaving utilities to face the problem of idled capacity, or stranded capital assets that are not generating any sales.

But although utilities might be reluctant to accept renewable energy, the security and economic arguments for solar are a powerful force. Japan now has one of the world's most generous subsidies for solar power—more than three times Germany's level. These subsidies, known as feed-in tariffs, pay producers of solar power a set amount for the electricity they sell to the electrical grid. Following the announcement of the feed-in tariffs, many Japanese companies developed an appetite for solar projects. One of them, the finance company Orix, is profiled in the following section.

~

Japan is famous for its plethora of little-used airports, built during the boom years of the 1980s and even the lost decade of the 1990s. This boom saw a huge number of roads, bridges, and other cash-hungry projects built, providing business for construction companies that in turn provided significant funding for the ruling Liberal Democratic Party. Today, the Japanese landscape is dotted with these airports—airports that are of no economic value as the rural population dwindles in a country that is served by a good rail system.

Orix, Japan's largest commercial finance company, has found a novel use for one of these little-used airfields at the southern tip of the southern island of Kyushu.[37] In March 2013, the island's small city of Makurazaki—best known for producing the fish flakes used in miso soup and home to a scant 25,000 people at the southernmost end of the JR rail line—announced that Orix would convert Makurazaki's municipal airstrip into an 8.2 MW solar farm.

On September 1, 2014, power started flowing from this unique solar farm. With the runway swathed in neatly edged solar panels, it looks more like a piece of installation art than a power plant, with the sharp-edged and coolly luminescent blue-black panels set off against the soft fertile green of surrounding rice fields.

This is not the only novel place that Orix has identified as a good site for solar power. At the other end of the country, on the northern island of Hokkaido, it is building a solar farm around a motor speedway, with the 21 MW facility expected to provide enough power for more than 6,600 households after it goes into operation. In between, the company is scouring the land-starved island-nation for suitable sites, most often leasing unused roof space.

In 2012, Orix said that it would invest about $1.14 billion in solar power by 2015, instantly catapulting itself into the ranks of a significant global player. Yet no one could mistake the company for a mission-driven group of environmentalists trying to run a socially oriented business. At its founding as the Orient Leasing Co. in 1964, the firm—one of Japan's first leasing companies—focused on the ship, aircraft, and heavy equipment industries, later diversifying into consumer and commercial finance, investment banking, and asset management. Orix is in a wide variety of businesses; in 2013, it acquired the Hyatt Regency Hotel in Kyoto from Morgan Stanley and bought control of a Mongolian bank. Its 2013 purchase of Robeco, a Dutch investment management company, for €1.9 billion is the largest in the company's history. Today, it has a presence in thirty-four countries and, like many Japanese companies, owns a baseball team—the Orix Buffaloes.

Orix's business is both elementary—borrow more cheaply than the rate at which its assets earn—and mind-numbingly complex, requiring the skill to weigh risks ranging from possible changes in government policies to the likely impact of a natural disaster to managerial competence. The Tokyo-based company, whose shares also trade on the NYSE, is sizeable: it reported $13 billion in revenues and net income of $1.8 billion for the fiscal year ending in March 2014. In September 2014, investors valued the company at $20 billion.

The idea behind Orix's foray into solar power is simple: the company leases space—whether a candy manufacturer's rooftop, the grounds at a motor speedway, or an abandoned airport—and puts up solar panels. It then sells the power to utilities at a very generous rate, guaranteed for twenty years through Japan's new tariff. The tariff rate is fixed according to the year in which the deal is struck but is being progressively reduced. It fell from ¥40 (40 cents) in 2012 to ¥36 (36 cents) in 2013.[38] As a result, companies have every incentive to get deals in place in order get a higher tariff. Although governments can and do change policies, Orix executives think that this is unlikely. "We think the risk of the Japanese government changing policies is low compared with European countries," says Atsushi

An 8 MW solar power plant built by Orix Corporation on the site of the former Makurazaki Airport on Kyushu Island, Japan that opened September 1, 2014. Photo credit: Orix Corporation

Yoshihiko Miyauchi, then chairman and chief executive officer, Orix Corporation, participates in a session at the World Economic Forum in Davos, Switzerland, in January 2009. Photo credit: © AP/Alessandro Della Bella

Murakami, a senior manager in Orix's business development department. "We have less political risk."

Yoshihiko Miyauchi, Orix's founder, is a genial, avuncular executive. Born in 1935, he is part of the remarkable generation of Japanese who were raised amidst the war and poverty of the 1930s and 1940s and who went on to rebuild Japan in the decades after World War II. Even in his late seventies, Miyauchi is an energetic traveler and is as at ease seated on the floor of a Burmese monastery for a meal of simple curry as he is eating elegant *kaiseki* cuisine in Kyoto. His cheery manner aside, Miyauchi is a hard-headed businessman who before the March 11, 2011, earthquake regarded renewable energy as a marginal business, notwithstanding small investments in areas such as waste-processing services and metals recycling. That all changed after the disaster at the Fukushima nuclear plant.

Orix is one of many companies that have responded quickly to the incentives. It plans to invest about ¥90 billion ($900 million) to set up 300 MW of solar-farm capacity around Japan. Separately, it plans to invest another ¥24 billion ($240 million) on 100 MW of large-scale rooftop solar facilities. In September 2013, it signed an agreement with discount retailer PLANT to use the roofs at two of the company's stores for solar installations designed to produce 2.7 MW—enough power for about 730 households. Orix has plenty of competition: SoftBank, Sharp, Kyocera, Panasonic, Marubeni, and Mitsui are among those that have also gotten into solar.

The goal for Orix is to invest in 400 MW of solar power over three years. (By way of comparison, the NASDAQ-listed U.S. solar installation and leasing company Solar City installed 280 MW in 2013.) For Miyauchi, the relative maturity of the technology, combined with a twenty-year warranty on the solar panels and a twenty-year fixed price for the power, makes this an easy investment decision. "Feed-in-tariffs are very favorable for investors. The solar price is fixed. The price is guaranteed by government. Why not invest? It's a very easy decision." The decision is even easier thanks to annual returns of over 10 percent, says Miyauchi.

The broader field of energy management has become an area for experimentation by Orix. In mid-2013, it established ONE Energy Corp., a joint venture with electronics giant NEC and EPCO, to provide household energy services management with a solution that includes solar power and rental storage batteries. It has also acquired a home energy management business that sells solar systems to homeowners and leases them storage batteries, providing backup power in case of electricity outage (and when the sun is not cooperating).

Other renewables areas are of interest to Orix as well. In 2011, a 13.6 MW biomass operation powered by wood chips went into operation. In Yokohama, where one of Japan's more interesting experiments on green cities is under way, the company is buying bulk power and selling it to residential condominium owners at a discount. In late 2013, it bought an 80 percent stake in the San Francisco–based energy services company Enovity.

Already operating a small geothermal project, Orix is scouting for a larger-scale test geothermal site.[39] "My idea is to invest more in geothermal," says Miyauchi. "With geothermal you need some level of sophistication. Solar is simple. Everyone is doing it." The company's renewable-energy business "is very small but it has a big future. If you have expertise, the returns can be good. I'd like to do more green energy—that is my dream."

Orix is emblematic in its search to grow businesses in the environmental services areas. Since its founding in 1964, it has cycled through a variety of businesses, from the heavy industry of the 1960s to the more consumer-focused businesses of today. Environmental services, whether solar panels or battery storage devices for consumers, provide new opportunities for profit from the difference between the company's low cost of capital and long-term investment horizon, on the one hand, and the situations of businesses and consumers who do not have the same access to capital, on the other. Miyauchi's recent enthusiasm for solar power in particular reflects the ways in which government policies can drive markets and the speed with which businesses can adapt to changing policies and changing market.

~

Like a shooting star, Suntech seemed to come from nowhere, whizzed across the sky, and disappeared in a fizzle of recriminations and lawsuits. The evanescence of its success underscores the challenges facing solar power. Measured by conventional business metrics, the industry is a failure. It has destroyed stock market value. Profitability is elusive. Even the industry's contribution to environmental sustainability has been called into question, thanks to revelations of dirty manufacturing processes and hazardous waste discharges.[40]

Yet to leave it at this is to misunderstand the full scope of the revolution. Solar power is an increasingly important part of the global energy mix, thanks to the success of Suntech and others in bringing prices down so far, so fast—with the result that solar is competitive with many other forms of conventional fossil-fueled power.

China's solar companies will almost certainly remain important over the next decade or more. Its strong manufacturing base, coupled with its large domestic market, ensures that China will play a central role in the solar industry. Profits may always be marginal because much of the solar industry is a low-margin commodity business. But the determined entrepreneurs and pragmatic, results-oriented national and local governments will do what they can to nurture the industry. This state support for privately run, capital-intensive infant industries is a model that South Korea used particularly successfully in shipbuilding and autos. As with solar today, South Korea's heavy industries lurched from success to shock, again and again, with companies negotiating for government help to get them through periods of crisis. One key factor in South Korea's success is the use of export sales as an objective marker to evaluate whether or not a company is worth saving. Another is the government's willingness from time to time to let companies fail.

Solar isn't just about manufacturing. That is an increasingly low-margin commodity business. DRAMs went from being a source of angst and trade disputes to being a capital-intensive and largely low-margin product, and the higher-value-added design-intensive chips and products became the focus of attention, and solar will experience much the same thing. What matters isn't making solar panels—rather, it is producing electricity from solar reliably and at a competitive cost.

China's ability to innovate technologically is at the center of the debate as to whether its high growth will continue. For solar power, the question is different. The issue is whether China, having driven down costs so dramatically and seemingly having so little concern about profitability, will smother new technologies. China could find itself outflanked on the technological front. The most important technological rival to standard photovoltaic panels is so-called thin-film solar, a technology that is currently more efficient and less expensive to manufacture. Some technologically adept rivals are in thin-film, including the world's largest solar company, Arizona-based First Solar. Another rival to watch is Taiwan's TSMC, one of the world's most important semiconductor manufacturers. TSMC in 2013 invested more than $250 million in a 700,000-square-foot manufacturing facility with an annual capacity of 300 MW of thin-film solar. The company said that the facility could be increased to produce 1 GW of thin-film solar panels by 2015.[41]

Although Chinese manufacturers dominate the industry for now, they will have to work hard to stay on top. Most important, they will need to find a business model that allows them to generate the profits needed for continued investments. For now, Chinese solar companies have shown that they can disrupt existing businesses, but they have yet to prove they can build long-lived ones of their own.

Suntech exemplifies these challenges. Its ability to raise money, from Chinese banks and international investors alike, to fund its fearless expansion program transformed the solar industry and for the first time made it broadly possible for solar power to compete against more traditional fuels. Yet the company's strength was its weakness, for its relentless expansion strategy left it fatally exposed when the business shifted in unexpected ways. Suntech's Shi Zhengrong was a revolutionary. But he was not able to reap the rewards of the changes he wrought, changes whose effects will be felt for many decades to come. The Sun King is dead, but his revolution lives on.

2

Blowin' in the Wind

> As yet, the wind is an untamed, and unharnessed force; and quite possibly one of the greatest discoveries hereafter to be made, will be the taming, and harnessing of the wind.
>
> —ABRAHAM LINCOLN

The hypnotic grace of wind turbines makes them the ubiquitous icons of a fossil-fuel-free future, the photogenic symbols of the clean-energy revolution. The thin, elegant towers, their clean, white lines capped with slowly twirling pinwheel blades, have the playfulness of a toy coupled with the technological brilliance of a new electricity source.

Wind has become a serious part of the energy mix. In Denmark, wind accounted for more than 30 percent of the country's electricity in 2012, and the country hopes to increase this to 50 percent by 2020. Wind produces 20 percent of Portugal's electricity and 18 percent of Spain's. In the United States, wind power generates enough electricity to power over 11 million households. Overall, there were 225,000 wind turbines producing power at the end of 2012, generating about 2.6 percent of global electricity. The growth of wind power has accelerated in recent years. In the five years from 2008 until 2012, wind power installations more than doubled, approaching 300 GW of installed capacity. That is a lot of power—roughly equal to the total electricity-generating capacity of Germany and France combined.[1]

Despite its growing prominence, wind power's importance in the overall energy mix is easy to overlook. Each individual turbine produces relatively little electricity, usually about 2.5 MW or so (although average generating capacity has been rising). A typical coal-fired power plant produces about

200 times that amount. But a wind farm with hundreds and hundreds of turbines adds up to a lot of electricity.

Indeed, wind is likely to be the most important source of renewable energy, excluding hydropower, over the next few decades. In 2013, the International Energy Agency (IEA) raised its 2050 target for wind's percentage of total global electricity production from 15 percent to 18 percent; as recently as 2009, when the IEA did its initial assessment of the wind market, the 2050 forecast was only 12 percent. The rapid spread of wind power, notes IEA Executive Director Maria van der Hoeven, is "a rare good news story in the deployment of low-carbon technology deployment . . . In some countries it accounts for 15 percent to 30 percent of total electricity production." Most important, in many places wind can hold its own without subsidies.[2]

Wind power, like solar power, relies on natural forces to produce electricity. Unlike fossil fuels, wind does not produce air pollution or carbon dioxide emissions. Its pricing is not subject to the same sort of fluctuations that afflict fossil-fuel prices, which reflect the changing prices of coal, oil, and natural gas. Its promise as a clean, economical fuel is unmatched in many locations around the world. Substantially cheaper than solar and typically cost-competitive with all but the cheapest fossil fuel in any given market (e.g., coal in China, gas in the United States), wind power has room to grow as wind turbine costs continue to fall. Technological advances, together with a growing demand for turbines, have brought about a dramatic reduction in the price of wind power worldwide and have made it competitive far more quickly than anticipated a decade ago. Prices already fell 25 percent from the end of 2009 through the end of 2013, according to Bloomberg New Energy Finance. As of late 2013, the cost of an installed wind turbine was about €900,000 a megawatt, cheaper than most competing fuel sources.[3]

This era of inexpensive wind power is the result of a new generation of higher-capacity turbines that can produce substantially more electricity than their predecessors. Taller wind towers and longer turbine blades are increasing the average power produced by each turbine, further reducing the cost of producing electricity. These are not the small, eggbeater turbines of the 1980s: current-generation turbines typically have blade diameters of 110 meters (360 feet) and sometimes even more—about the distance from home plate to deep center field in Boston's Fenway Park.[4] Simply put, these are monster blades. Just moving a single blade from factory to wind farm is a piece of logistical engineering.

The economics of wind are so compelling that growth in new turbine installations will remain strong for many years. In 2014, the Global Wind

Energy Council estimated that 278 GW of wind power would be installed from 2014 through 2018. That would mean almost a doubling of the world's 2013 installed base of wind turbines.[5] China and the United States will remain the world's largest individual markets. Europe and India will also remain significant sources of demand. Over the next decade or more, new government policies to cut emissions, combined with the improving efficiency of wind turbines, are expected to support a growth trend for wind energy. Longer term, new energy storage options currently in development—large-scale batteries, for instance—will likely enable the share of wind energy in power markets to grow even more. The lack of storage has long been one of the great weaknesses of wind and solar, making them intermittent, unreliable sources of energy. Research into cost-efficient storage systems could someday allow wind and solar to be counted as part of the baseload power on which utilities rely.

Countries with good wind resources, led by China, will continue to grow their installed base of wind power generation. However, wind power's days of torrid growth—some 22 percent annually in the ten years to 2012, or a doubling about every three years—are over, as the simple arithmetic of a larger installed base makes it mathematically difficult to maintain that sort of growth percentage. Moreover, the European subsidies that, coupled with falling turbine prices, spurred the industry during the recent past are largely gone. Yet the economics of wind power and the large wind resources in the world's two largest economies mean that the absolute amount of wind power installed will continue to show impressive growth over the next decade. The IEA predicts that by 2020 China alone will overtake all of developed Europe (defined as those nations belonging to the Organisation for Economic Co-operation and Development) as the leading producer of wind power.[6] Given China's penchant for periodically ratcheting up its environmental targets, including planned wind installation capacity, it is more likely to exceed than to fall short of the IEA forecast.

The speed with which China embraces wind power could be the single most important factor in ending its dependence on coal. This transition can happen much faster than most people realize. A joint study by Michael McElroy and others at Harvard University and Beijing's Tsinghua University found that *all* of China's additional electrical power needs theoretically could be met by wind by 2030 at a price comparable to that of coal power. The speed at which China adopts clean energy is, above all, a political decision.[7]

China's emergence as a major player in the industry is a very recent development. In 2000, China was on the extreme periphery of the global wind power industry. By 2010, it had a larger base of wind turbines than any other country, and a handful of its companies had emerged as among the world's largest. Wind energy is a tale that has captured both hopes and fears, in China and around the world. There is hope that China will succeed in embracing clean energy for the sake of its ashen skies and its people's blackened lungs, but there is also fear that its cash-rich, government-backed wind companies will slash prices and thus obliterate foreign competitors and their green jobs.[8]

~

At the turn of the century, it would have been an exaggeration to say that China's wind turbine makers made up an infant industry. It was not until 2000 that the first turbine built by a Sino-foreign joint venture was installed—China's few existing turbines had been imported. From that humble beginning, the industry shot up. The next year the first wind turbine with Chinese intellectual property content was installed. In 2008 came the first Chinese turbine with 1 MW of capacity, spurred by a government program paying a cash prize to manufacturers of the first fifty 1 MW capacity turbines. Today, China is the only country building gigawatt-scale wind power plants—massive wind farms that have the capacity to generate as much electricity as two average coal-fired power plants. Chinese companies now have about 20 percent of the world market in wind power.[9]

Government policy has been indispensable for, and inseparable from, the success of China's wind industry. The story with wind is the same as that of so many other industries in which Chinese industrial policy has played a key role. A process of trial-and-error experimentation, over a period of more than three decades, resulted in China's attaining global scale in the industry. Yet questions remain about the country's ability to achieve top-tier technological capabilities and to be able to compete in the top ranks of international players.

China started experimenting with wind power as early as 1986. The Dabancheng wind farm in Xinjiang—the country's first—went into operation in 1989. This experimentation with wind power was a cautious, stop-start process, whose only tangible result was the installation of a small number of imported turbines. China tried to make its own turbines, but by the mid-1990s, this effort had largely failed.

However, over the course of that decade, favorable government policies helped spur growth. In 1994, the Ministry of Electric Power ordered grid

Workers from Xinjiang Goldwind Science and Technology Co. install a 3 MW wind turbine generator system at the Dabancheng wind farm in Urumqi in northwest China's Xinjiang Uygur Autonomous Region on December 10, 2009. Photo credit: © AP/Zhou You/ColorChinaPhoto

operators to buy wind power despite its higher cost, although the controversial and unsuccessful program was eventually halted. The Ninth Five-Year Plan (1996–2000) provided modest funding for renewable energy, including wind turbines with 40 percent local content, meaning that 40 percent of the turbines' components had to be manufactured in China. At the same time, a loan program offered wind farms bank loans at half the price of the benchmark commercial rate, with preference given to wind farms that used turbines with a higher local content. Several small demonstration projects resulted.

The late 1990s and early 2000s saw a dramatic uptick in policies—some successful, others not, but together acting to turbocharge the industry. Under the Ride the Wind program initiated in 1997, foreign wind industry leaders like Vestas, Gamesa, and General Electric were invited to create joint ventures with Chinese partners. These foreign companies were asked to share technology and expertise in exchange for potential access to China's vast market. Ride the Wind had a local content requirement of 20 percent. That same year the Double Increase program was inaugurated, with the goal of doubling the wind turbine capacity through low- or no-interest loans and a focus on domestic manufacturing. In 1998, import duties on wind turbines were modified, explicitly encouraging imported components, while financially penalizing the importation of finished turbines.

Altogether, this cluster of new policies led, in 2000, to the installation of the first wind turbine manufactured by a Sino-foreign joint venture. Following this success, Chinese companies started to set up their own manufacturing programs, typically licensing foreign technology. The first decade of the century saw additional government policies designed to jump-start the local wind industry. Strong local content rules discriminated against foreign producers, helping ensure that local manufacturers who expanded would have a market. What happened in wind paralleled efforts in a number of Chinese industries, such as automobiles, high-speed trains, and nuclear power (to name just a few of many), where China pursued a focused policy of bringing more and more of the manufacturing know-how to local companies. Absorbing foreign technology is a process that many countries have attempted, but few have been as successful at it as China and its neighbors Japan, Korea, and Taiwan. The results in terms of the scale and sophistication of China's wind turbine production, as in so many other areas, have been dramatic.

There was continued research and development spending throughout the early 2000s, focused on turbine localization. Value-added taxes for wind-generated electricity were cut from 17 percent to 8.5 percent. A program to build wind farms was implemented, with an initial 50 percent minimum local content requirement in 2003 rising to 70 percent for these projects in 2005. Also in 2005, China's State Council, the country's highest administrative body, set up guidelines for Chinese wind farms to receive subsidies under the Clean Development Mechanism, a financing scheme set up under the Kyoto Protocol to encourage the development of low-carbon projects in the developing world. In succeeding years, similar sorts of support proliferated.[10]

Slowly, Chinese engineers acquired more expertise. Government policies nurtured the companies and, at critical moments, protected them against foreign competition. In a push to increase the installed base of wind turbines, wind concessions were granted—none went to foreign operators—and government mandated that utilities and grid operators buy the power. Massive wind farms were built, and China's installed base doubled from 2005 to 2006.

The Renewable Energy Support Law of 2006 provided key support for an industry that by then was strong enough to take fuller advantage of government help. The law required power grid operators to purchase specified amounts of electricity from registered renewable-energy producers. The law also offered financial incentives, such as a national fund to foster renewable-energy development, inexpensive bank loans, and tax preferences for renewable-energy projects. The growth rate of China's wind industry accelerated, encouraged by these government incentives. As a result, total Chinese installed capacity doubled again from 2006 to 2007, to 5.9 GW.

Significantly, 2006 was also the first year that China adopted an energy intensity target, as part of the country's Eleventh Five-Year Plan (2006–10). A seminal speech in mid-2007 by President Hu Jintao, committing China to meet energy intensity and carbon intensity targets, provided a significant directional signal underscoring the importance of green energy to the country's development. President Hu's backing encouraged the further development of policies supporting wind and other clean-tech initiatives.

Thanks to government policies that favored Chinese companies, almost all the new business went to domestic manufacturers. In 2004, foreign manufacturers had a 78 percent market share of wind turbine installations, almost all of it belonging to Vestas of Denmark, Gamesa of Spain, and GE of the United States. By 2006, foreigners' share was down to 55 percent, and in 2009, it had dropped to 14 percent. Renewable energy, including wind, was a major beneficiary of China's mammoth $586 billion stimulus program, enacted in 2009 as a response to the financial crisis; that year saw the introduction of a feed-in tariff for wind power, providing an important subsidy.[11]

Coincidentally or not, in 2009 China installed more wind capacity in a single year than any other country. The following year China had the world's largest installed base for wind power, with 44.8 GW of wind turbine capacity, squeezing past the United States's 44.1 GW. This growth was accompanied by the dramatic expansion of Chinese wind turbine companies. In 2006, there was not a single Chinese wind turbine maker among the world's top ten. Yet by 2010, four Chinese companies—Sinovel, Goldwind, Dongfang Electric, and United Power—ranked among the ten largest wind turbine manufacturers.[12]

Vestas, the world's largest and one of the oldest wind turbine manufacturers, felt the brunt of China's rise. In 1986, Vestas had sold China its first wind turbines.[13] Twenty-five years later, Vestas was fighting for its corporate life. By 2012, amidst a series of management reshuffles, restructuring, and brutal cost-cutting, a Danish newspaper reported that Xinjiang Goldwind Energy Company, China's largest wind turbine maker (and one that we will discuss in detail below), was one of a pair of potential Chinese buyers

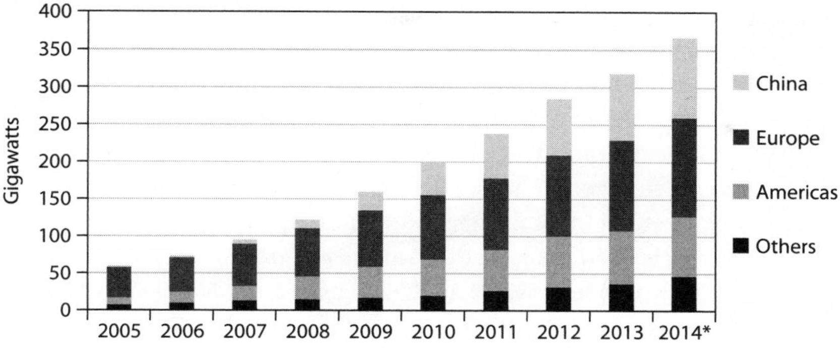

Global Wind Capacity
Wind has become a significant source of electrical power in China. Note that 2014 is estimated. Source: BP Statistical Review of World Energy and Global Wind Energy Council

looking to acquire the company. Although Vestas remains independent at the time of this writing, even the idea that Goldwind might be a bidder for Vestas was remarkable proof of the Chinese wind turbine industry's extraordinary rise.

By the end of 2013, China's installed base was 50 percent larger than that of the United States, and the gap is projected to widen substantially. China's base had grown thirty-fivefold just since 2006, from 2.6 to 91.4 GW—growth of a sort that perhaps only China is capable of. Still more growth is on the way. The Twelfth Five-Year Plan (2011–15) sets a target of 100 GW of installed wind capacity by 2015. Wind is now the third largest source of electricity in China, surpassing nuclear in 2012 and trailing only coal and hydropower. The State Council issued a *White Paper on China's Energy Policy 2012*, setting a target of 30 percent of electrical power generation from non-fossil-fuel sources by 2015, and wind is a significant component.[14]

Chinese wind turbine manufacturers have not eliminated the technological advantage enjoyed by companies such as Vestas, GE, and Siemens, and they have failed to match the solar makers' success in expanding outside their domestic market. Goldwind, China's largest and arguably most international turbine manufacturer, derived only 11% of its $1.9 billion in 2013 sales from abroad. It counts Australia, Panama, Cuba, and Romania as its largest international customers, countries whose installed base together comprise 2% of the global figure. However, Chinese companies benefit from a dominant position in their brutally competitive home market, with foreigners' 5 percent market share likely to fall to just 1 percent.

As with solar, China's rapid increase in wind turbine production is contributing to overcapacity and falling prices. The Chinese Wind Energy Equipment Association, a trade group, said that Chinese equipment prices fell 23 percent from 2009 until early 2014. Shen Dechang, the group's vice secretary-general, said that profits also had dropped and called for industry consolidation, predicting that the number of turbine makers could fall by two-thirds in the next five years. Shen's prediction is unlikely to come to pass, absent a searing economic crisis. From steel to cement to autos, industry association officials, academics and other experts have since the turn of the century been outspoken in calls for consolidation of various industries. However, as discussed in the solar chapter, China's under-priced capital and land and a continuing abundance of reasonably priced skilled labor means that its economy is dependent on investment to a degree that is utterly without historic precedent. With local government officials still measured by their ability to promote economic growth, and continuing macro-economic incentives that argue in favor of expansion, near-term consolidation of the sort that Shen predicts is unlikely.[15]

A proprietary Bloomberg index tracking the world's leading wind turbine makers calculated that the industry lost almost half a billion dollars ($461 million) in 2012. This is in stark contrast to 2011, when the industry registered $5.25 billion of operating profits. Profits improved in 2013 to $488 million, but were still far short of the 2011 peak.[16]

Falling profitability is not the only issue facing the nascent Chinese industry and its overseas competitors. Another is the difficulty of moving from installation to usable power. In 2013, China built more than 16 GW of new wind capacity, but it is better at installing wind turbines than generating electricity from them. Although China has much more wind capacity than the United States, the latter actually generates more usable electricity from its wind turbines. Wind produced only 2.6 percent of China's electricity in 2013, although wind turbines accounted for 5 percent of installed capacity, suggesting that as much as 60 percent of the potential capacity is being squandered. The Global Wind Energy Council estimates that in 2011 this wasted capacity was equal to about 10,000 GW hours of electricity, about the same amount as the total electricity produced by Hong Kong or Singapore in an entire year.[17]

Why is this waste occurring? The answer reflects some of the weaknesses of China's top-down economy. A central government decision to dramatically increase wind power was made without thinking through the detailed technical implementation. Turbines were built but not promptly hooked up to the electrical grid because the necessary grid capacity did not always extend to the remote areas where wind farms have traditionally been located.

The problem does not stop at grid connections. Even when turbines are connected to the grid, China's grid operators do not always use the electricity they produce. Some of this reflects technical issues with the grid, but there are also issues of favoritism and corporate politics between the grid operators and wind farm owners. Even when the power is used, grid operators sometimes do not pay the wind companies on a timely basis, a situation that occurs even between two companies within the same corporate group.

Exacerbating matters are rigid price controls. Wind is more expensive than coal. Grid operators and power companies do not like to pay for more expensive wind, and often do so grudgingly, because they are unable to pass these costs on to their customers. The Chinese government was trying to jump-start wind power by promising producers that they would get a higher price for that power. But rather than paying the subsidy itself or allowing companies to charge consumers more, it pushed the expense onto companies.

The problems in hooking wind up to the grid and in effectively dispatching the power were so severe that former Premier Wen Jiabao singled these issues out in a major national speech shortly before leaving office in 2013. These delays are perhaps surprising, given that state-owned enterprises account for almost 90 percent of all wind farms, that the grid itself is state-owned, and that all electricity producers are also state-owned.[18]

The situation reflects China's often-conflicting policies and shows some of the limits of its state-directed capitalism. It wants cheap electricity to power its economy, and that means coal, which accounts for about four fifths of its electricity production. Yet China wants to lessen its reliance on coal by nurturing renewable energy. Wind farms were often built at the behest of the government, sometimes as a way for companies to show their support for the policy and to curry favor.

China's ability to generate usable wind-powered electricity will be a test of its "learning by doing" approach to environmental policy—indeed of its overall approach to economic reform. New government directives now in place call for significant investment to build out the power grid and for new supply to be approved and tested before going online.

China's newfound dominance in wind turbine manufacturing reflects a remarkable story, although to readers familiar with China, the story has many familiar contours. Wind turbine manufacturing developed rapidly, from an embryonic industry that relied on foreign technical help, through an adolescent period when the government prodded and cajoled companies to increase their technological sophistication, to today's position, a sort of early adulthood in which the industry is the world's largest but is not yet fully mature and stable—as shown by the lack of grid connections, the failure to use power from wind turbines, and the lagging payments. Still, whatever the systemic barriers in China, many remarkable companies have emerged. These success stories often start with a tenacious leader who is able to demonstrate entrepreneurialism in a Chinese context—entrepreneurialism that seizes market opportunities but simultaneously nurtures the government relationships that are critical to success in China. One of the most striking examples in renewable energy is Xinjiang Goldwind Science & Technology Co., whose founder is a scientist determined to stay away from politics and the government but now finds himself enmeshed in both as the head of China's largest wind turbine maker.

~

The story of Goldwind and its founder, Wu Gang, is remarkable even in a country filled with extraordinary tales.[19] Wu, a 1983 graduate of Urumqi's Xinjiang Engineering Institute, was born in 1958, making him a teenager during the upheaval of China's Cultural Revolution. Like so many other young people at the time, he was sent to spend two years in a remote village, the last six months of which he spent teaching English to middle school students just a few years younger than he was. Wu worked late into the night on his own English studies and the next day taught what he'd just learned—an experience that even three decades later he remembers as nerve-wracking. His English skills were crucial to him in his later work with Goldwind, allowing him to get the most out of visits by foreign scholars and wind industry representatives.

Wu's journey toward Goldwind started while he was teaching at a technical college in his hometown of Urumqi and suddenly had, as he says, "a sense of crisis." Wu felt that if he continued teaching, he would squander his life. "I had to find a new challenge," he remembers. It was the mid-1980s, and the excitement of the Deng Xiaoping reforms was sweeping China. "My major was power, and I was looking for the future energy source." As an engineer, he says, "I focused on technology."

He wasn't the only one with this focus. In 1985, the Xinjiang provincial government's wind research program teamed up with the country's

Ministry of Water Resources to organize a delegation to Britain, Germany, and Denmark. In one of those serendipitous events that could have been a boondoggle but turned out to be a brilliant inspiration, the delegation decided, on the spot, to buy a Danish wind turbine to bring back to Urumqi, the capital of Xinjiang.[20]

Few cities in China are deeper in the interior, or more remote physically and culturally, than Urumqi. Midway between the Iranian capital of Tehran and the Chinese capital of Beijing, Xinjiang is where China becomes Central Asia. It is an almost unimaginably harsh region, one where the Gobi and Taklimakan Deserts meet. Muslim Uighurs outnumber Han Chinese, although Xinjiang does not have many people. What Xinjiang has, in abundance, is wind.

Xinjiang Wind Energy, one of China's pioneering wind companies and Goldwind's predecessor, was founded by using a $3.2 million Danish grant.[21] The company was led by Wang Wenqi, a stubborn man who believed that wind could be harnessed for electrical power. Wang, who at the time was also the director of Xinjiang's Irrigation Works and Hydropower Research Institute, remembers that colleagues scoffed at his vision of wind power as something "only creatures on the moon could think of." But the idea that energy could be produced by wind captivated Wu Gang, and he joined the fledgling company in December 1987. Together, Wang, Wu, and their team persisted in their vision of wind power and built China's first wind farm at Dabancheng.

Wang Wenqi imbued his team with a sense of mission. Wu Gang remembers doing the fieldwork necessary to decide where to site this first farm and describes traveling around the province in a Beijing Jeep, itself a product of one of the first Sino-foreign joint ventures. The researchers took turns standing outside in the wind for twenty or thirty minutes at a time, holding wind-measuring equipment in temperatures as cold as −20 degrees Celsius (−4 degrees Fahrenheit) before heading for the shelter of the Jeep and turning the equipment over to another colleague. Dabancheng had fairly constant wind—enough but not too much, with fairly even distribution. It proved to be an ideal site for a demonstration project showing wind power's potential in China.

Wang Wenqi's vision to build a wind farm at Dabancheng was a remarkable act of foresight. Now, when China's prowess in everything from steel to satellites is taken for granted, it is hard to remember how poor and how technologically underdeveloped China was in the mid-1980s. Deng Xiaoping's economic reforms had been launched in 1978, but despite a flurry of excitement at the prospect of change, tangible progress was limited. Black-and-white televisions and electric rice cookers were aspirational goods, totemic symbols of affluence. A fixed-line telephone was beyond hope for most people, private cars unthinkable. The country's backwardness was magnified when one moved away from the coasts and from Beijing to interior cities like Urumqi.

When it opened in 1989, the Dabancheng wind farm consisted of the thirteen small (150 kilowatt, or kW) Danish-made Bonus turbines.[22] The farm in turn seeded China's industry, with engineers and others from around the country visiting on study tours in the years after it opened. Many of those—like Wu Gang, who headed the Dabancheng project—later fanned out around the country to install wind turbines and, at least in Wu's case, to start their own wind power companies.

Wu and a group of colleagues founded Goldwind in 1998, as a successor company to Xinjiang Wind Energy Company. The goal was "establishing Xinjiang as the birthplace of China's domestic wind energy industry." Fast-forward to 2013, and Wu's Goldwind was China's largest manufacturer of wind turbines and the world's second largest, with

Goldwind turbine in Guanting, near the Badaling section of the Great Wall, northwest of Beijing. Photo credit: Goldwind

a 10 percent global market share that year. It had installed more than 14,000 turbines, on every continent except Antarctica, capable of producing 19,000 MW of electricity, more than the entire installed generating capacity of Switzerland.[23]

How did Goldwind grow so big so fast in an industry where China had no expertise only twenty-five years earlier? The company's success reflects a combination of traits often seen in China's best companies: strong engineering skills, an ability both to acquire technology from other companies and to develop its own technology, and political savvy in its industry. That the wind industry is enjoying a period of rapid growth, both in China and globally, has provided a critical boost. So has the backing of Chinese national policies. A closer look at the role each of these factors has played in Goldwind's success provides a good overview of how many modern Chinese companies have developed.

Goldwind is a company dominated by engineers, reflecting China's deep strength in this area. Guo Jian, who stepped down as president and executive director in 2012, also holds an engineering degree from the Xinjiang Engineering Institute, and both Wu and Guo also have master's degrees in engineering. They and their team have been involved in numerous central government–led scientific research projects—including the 863 research program, the country's umbrella high-technology research program—and have won numerous awards. Wu has led a number of scientific and technological projects, including the Key Technologies R&D Program, that are part of China's Five-Year Plans.

This technical expertise means that it has been relatively easy for Goldwind to catch up in an industry as young and fast-changing as the wind power industry has been over the past twenty-five years. China aggressively bought foreign technology to help jump-start its wind industry, and the engineering skills of companies like Goldwind allowed this technology to be assimilated efficiently.

From Xinjiang Wind Energy's founding, licensing foreign technology was key. In 1996, the company licensed a turbine design from Jacobs Energie, a small German manufacturer. This policy of technological acquisition through licenses was mirrored by Xinjiang Wind Energy's competitors and reflected a national commitment to become a significant force in wind turbine manufacturing. The licensing strategy continued through several generations of turbines. After a merger in 2001, Jacobs became part of REpower Systems Group; that same year Wu's company, newly re-named Goldwind, licensed a larger turbine from REpower.

In 2003, Goldwind began what has become a long-running and successful partnership with German turbine designer Vensys Energiesysteme, licensing a yet-larger turbine from Vensys. Jacobs and REpower were manufacturers, but Vensys is a design house, eliminating the possibility that the Chinese company would ever compete with its German partner for sales. The partnership was so successful that Goldwind bought 70 percent of Vensys in 2008 for €41.24 million.

Although Vensys is a subsidiary of Goldwind and Wu Gang chairs the company, the German design firm continues to sell to other customers around the world. This international experience allows Vensys to give Goldwind insight into global trends in the industry. The link between the two companies is further institutionalized through Jürgen Rinck, who serves as Vensys CEO and Goldwind vice president and chief technology officer. So although Chinese wind companies have yet to acquire a major wind turbine producer such as Vestas, they have been successful in acquiring companies with strong underlying intellectual property.

Engineering expertise and smart licensing and acquisition decisions have been crucial to Goldwind's accomplishments. But to fully understand the company and the reasons for its success, it is important to appreciate the central and provincial government backing it enjoys. Although the company is publicly traded on both the Hong Kong and the Shenzhen exchanges, its largest shareholders are state-owned companies. Its major shareholder is China Three Gorges (operator of the eponymous hydropower dam), which is itself wholly owned by the central government's State-owned Assets Supervision and Administration Commission (SASAC).

SASAC oversees the central government's stakes in more than one hundred companies, including the country's dominant petroleum, utility, and telecommunications companies. Its shareholdings are worth hundreds of billions of dollars. At the time of Goldwind's Hong Kong initial public offering in 2010, its trio of executive directors included Wu Gang, Guo Jian, and Wei Hongliang, a decade-long employee of the China Three Gorges group who was responsible for capital management and investment. Central government support complemented the active backing of the Xinjiang provincial government, which Wu had enjoyed since the founding of the company in 1998.

Wu appreciates the irony of Goldwind's close involvement with the government. When he was younger, perhaps in reaction to the hyperpoliticization of the Cultural Revolution, he promised himself that he would stay out of politics and never become a government official. His technical and scientific background should have ensured this. But government support, first for wind research and later for Wu's company, meant that his role inevitably became more political. As wind, and the company, became more successful, ties with government became stronger, as symbolized by a 2006 headquarters move from Urumqi to Beijing. Gesturing around his office set in a substantial building on a large campus outside of Beijing, Wu says, "I promised that I never would be involved with politics. Now, unfortunately, all this I have accepted."

Government backing also is apparent in a $6 billion line of credit pledged to Goldwind by the China Development Bank, the country's policy bank, and specifically earmarked to support the company's goal of getting 30 percent of its revenue through international sales by 2015. Although Wu says that his company will increasingly work with international banks, China Development Bank's pledge could be crucial as the young manufacturer tries to win contracts abroad. This cash will be especially useful, given that customers increasingly want Chinese wind companies not just to manufacture turbines but to build the wind farms as well, pushing the construction risk—and additional costs—onto the Chinese firms.[24]

Goldwind's most prominent international project is a 110 MW wind farm in the heartland of the United States, in Illinois. The Shady Oaks project has a twenty-year electricity supply agreement with utility Commonwealth Edison. Goldwind broke ground on the $200 million project in September 2011 and announced that it had sold the project after its completion in 2013. Wu says that this strategy—taking the financial and project risk of managing the construction process and delivering an operating wind farm to the customer—may serve his company well in the future.

Goldwind's rise to the top of the industry has been difficult. It has moved to manufacture more components in China, at its own operations or with trusted suppliers, in a localization process that's had its challenges. The company was forced to undertake a massive recall after it encountered problems with the bearings in the turbine blades. The culprit was substandard steel used in the manufacturing process. The incident, writes wind energy expert Joanna Lewis, "illustrated how one small supply-chain error can result in a disastrous public relations problem for an emerging company in a highly competitive market."[25] However, the company is continuing to increase local production of parts. In 2010, it bought two companies in the Xiexin Wind Power group that design and manufacture turbine blades.[26]

Well-publicized problems like Goldwind's recall, which other Chinese manufacturers have also endured, have created a perception among foreign buyers that Chinese products may not last for the promised (and industry standard) twenty-five years. Wu bluntly acknowledges that wind is an immature industry and that quality needs to be continuously improved. "I saw a lot of failures," he remembers. "Towers fall over and people fall down, drop down dead." Wu says that his experience with fatal accidents has pushed him and the company to improve quality. "We accept we have distance with Siemens, Vestas, and GE," he says. "If you accept you have a distance, you have a power to improve. If we make mistakes, we don't argue with the customers. We say we are sorry, we are wrong, and we correct it immediately."

Given the uncertainty and the volatility in the industry, it would be premature to say that Goldwind's place as an industry leader is assured. If my cursory visit to one of its manufacturing facilities in November 2013 was any indication, the company has a long way to go to match its more advanced competitors' manufacturing standards. There was little activity among the shop floor workers—my guide explained that this was because budget targets for the year had been reached. But a general sense of disorganization and even untidiness prevailed, a feeling quite different from that experienced in well-run factories in Taiwan, Korea, Singapore, and Japan or in multinational operations in China.

Goldwind and its competitors are faced with the challenge of building a new generation of turbines for offshore wind projects, turbines that must be engineered to withstand even more extreme conditions than today's turbines. Still, Goldwind seems like a long-term winner due to its in-house technology, vertical integration throughout the supply chain (including a high degree of localization), solid manufacturing basis, global sales and service network, and increasing expertise in building and operating wind farms. The company's relationship with its German design-house subsidiary Vensys is unusual and should continue to give it a window on global trends that many Chinese companies lack.

~

China likes to say that it has pioneered a unique model of economic development, one that pairs the best of state-led planning—and state-backed financing—with the entrepreneurial instincts of individual companies.

In fact, there is little unique about China's path, except for its scale. China has largely replicated the Northeast Asian statist model pioneered by Japan and followed by South Korea. China's chosen industries, such as wind and solar, are sometimes different from Korea's and Japan's. But the basic model is the same. A repressed financial system, one that punishes savers and gives governments and companies access to large amounts of underpriced capital, is key. The government funds research at academic institutions, at state research bodies, and at companies that, even if nominally private, are in fact closely tied to the state. Successful companies are rewarded with inexpensive loans, more research funding, and more business. The general direction for these companies comes from the top, beginning with state leaders such as former President Hu Jintao and former Premier Wen Jiabao; this guidance is supplemented with more specific national laws and with administrative regulations developed at the ministry level, which in turn form the basis for detailed policies implemented both by ministries and by provincial and local governments.

This process is rarely linear and inevitably political. Although wind is an example of a fairly successful overall policy, there have been many false starts and missteps. China's pragmatism and its willingness to keep experimenting as it moves toward a goal—in this case, a technologically advanced, indigenous wind industry—have proven crucial to its success. The dependence of companies on government for many key aspects of business means that relationships with government officials are part of the process in a way that they would not be in most economies in the developed world. For example, access to financing (at below-market rates or not), land for expansion, and research funding all depend on government relations.

Wind differs in some important respects from solar. The wind turbine market for Chinese manufacturers remains mostly domestic and large-scale. Buyers are generally utilities and other state-owned enterprises. Solar is more dispersed and less capital-intensive—many people have rooftop solar panels, but few have rooftop wind turbines. Chinese wind turbine companies have only limited success overseas. Whether this is because foreign wind turbine buyers are more demanding, more risk-averse, or less price-sensitive is not clear, but the fact remains that Chinese turbine manufacturers have not had the same success as their solar peers in convincing overseas buyers that their products are as good as foreign competitors such as GE, Vestas, and Siemens.

There are fewer wind turbine makers than solar panel manufacturers, reflecting the higher capital costs in wind turbine manufacturing, and consolidation in the industry is likely to further reduce the number of Chinese turbine manufacturers. As a result of the stiff domestic content requirements, Chinese technological capabilities have increased dramatically in the past ten years. Yet, as in solar, China's ability to be a force for technological innovation remains in question. Harvard professor Michael McElroy estimates that U.S.-made wind turbines are substantially more efficient than Chinese ones, which convert only 21 percent of potential power into electricity versus 32 percent in the United States, with turbine quality accounting for an estimated 40 percent of the shortfall.[27]

Wind power has an extremely promising future in China, technological and policy limitations notwithstanding. The central government's backing for wind power, coupled with the country's rich wind resources, will ensure a robust domestic market. The only question is how quickly wind will grow and when it might displace coal. In 2012, the installed capacity of coal alone was just over 700 GW, about twelve times larger than that of wind. Yet this balance is shifting as the price of wind power falls and that of coal rises, especially as the environmental costs of coal begin to be included in calculating the price that China pays for its overwhelming reliance on the dirty black rock. The sort of rapid transition that is economically and technically possible will require an even more aggressive commitment by the Chinese government to move to a clean-energy future. But it bears repeating: the technology is here. The companies are ready. The economics are favorable. All that is lacking is full political commitment.

II

Our Human World: *Cities, Buildings, Wheels*

This is not about philanthropy.

—BRUNO LAFONT

More than half of the people on our planet live in cities, and the percentage, as well as the absolute number, is increasing rapidly. By the middle of the century, about two thirds of the human population, some 6 billion people, will live in cities—a number equivalent to the world's total population as recently as 2000. Asia and Africa are the places with the most dramatic urbanization. China already is home to the greatest urban migration in history, one that has seen its cities swell from 200 million in 1980 to 700 million today. By 2030, Chinese urbanites will number about 1 billion, more than two thirds of the country's total population of 1.4 billion. China's city dwellers alone are more than twice the population of any other single country except India.[1]

Cities can be dystopian nightmares, choking on their own waste—Petri dishes that culture disease and violence as well as tangles of pollution and congestion. Despite these problems, cities keep growing. Their lottery of human encounters promises opportunities that are found nowhere else: chances for poor immigrants from the country and elite professionals alike to advance their careers, to sample different experiences, and to enjoy better health care, better schools for their children, and for the lucky, the arts and culture that cities host.

If we can get our city lives right—especially our buildings and our transportation—we will be well on the way to meeting the challenge of environmental sustainability. More than ever, cities are where the people are, so sustainability efforts must focus on urban areas. Cities also offer efficiencies of scale, meaning that urban initiatives are likely to offer higher payback over shorter periods for investments in green buildings and more energy-efficient transit. "Better cities means resolving issues of housing, density, technology, infrastructure and aesthetics," says Bruno Lafont, the chairman and CEO of the world's largest cement company, the Lafarge Group, whose future depends on the growth of cities in Asia. For customers in countries like China and India, cost matters in a way that it does not for the rich world, where people have the money to pay a premium for more environmentally benign products. Lafarge and other leading cement and building materials companies cannot ignore the reality that they are competing in extremely price-sensitive markets, but they simultaneously are trying to take into account the long-term impact their products have on the environment and society. "How can we reduce the cost of construction when we take into account the life cycle of the building and not just the cost of construction?" asks Lafont. "On sustainability, how do we reduce the environmental footprint of the cities?"[2]

These are important questions—and maddeningly complex ones. Think of energy use as a three-legged stool. One leg of the stool is buildings; the energy used to heat, light, and cool buildings accounts for about one third of our total energy consumption. The second leg of the stool is transportation, with cars, trains, trucks, buses, and planes accounting for roughly one third more. The third leg of the stool is manufacturing, mining, and construction. Cities obviously figure prominently in the first two legs, but they are important for this final leg, too. Coal mining uses significant energy; coal in turn is largely used to generate electricity, most of which is used by city dwellers. More efficient buildings reduce the need for coal-fired power plants, and designing buildings to incorporate efficiencies and lengthen their useful lives cuts down on energy use by the construction sector. Each leg of the stool, in other words, affects the others, and the energy-saving opportunities available in a virtuous circle of more efficient buildings, transport, and city design are enormous. But this feedback cycle can work in the opposite direction as well, and large parts of Asia suffer from a vicious circle of inefficient buildings, transportation, and cities.

Asia also has a handful of exemplary cities that are consciously using their density as an advantage in ensuring an environment that is more pleasant as well as more environmentally and economically viable. Several Japanese cities, including Tokyo, have made good efforts. So, too, has Seoul. But the most striking example is the city-state of Singapore, which has pursued an integrated strategy to meet the challenge of a variety of interrelated issues centered on water and energy. It has established a clear track record that puts it among the global leaders in building a greener and more energy-efficient city. Thanks to an overall focus on sustainability by the nation's leaders, its urban design, building codes, and transportation policies have all been developed and implemented together, multiplying their impact.[3]

That Singapore has done this in a mere half-century since independence, which came at a time when the people of this poor nation struggled to survive, makes its accomplishments even more impressive. The small country's founding father, Lee Kuan Yew, who devotes a chapter of his memoirs to "Greening Singapore," faced lengthy resistance to his cleanup campaigns from Singaporeans who were too caught up with daily survival to worry about the environment. Initially, his was an environmental campaign based on cleanliness and civic order rather than on environmental sustainability. But as the extended discussion of Singapore will show, it has developed into a comprehensive, scientifically based urban sustainability program that encompasses a range of concerns from tree planting and botanical gardens, to water security and water sports, to deliberately high costs for automobiles, to ever-tighter green building requirements.

There are also many new cities that are being built—so-called greenfield projects that give visionary architects and urban planners the luxury of working on a largely unmarked canvas. Although recognizing that the battle for urban sustainability will largely be fought in existing cities, these new cities can serve as laboratories and pacesetters, and they deserve a close look.

Buildings are what define a city. They are also responsible for most of a city's energy consumption. It's particularly important that Asia's many new buildings be designed, constructed, and operated more efficiently. China alone is building more than half of all the floor space in the world, and not surprisingly, it also accounts for over half of global cement production. Today's buildings will last for decades; the more efficiently they are built and operated, the more savings will accrue. Conversely, bad buildings will be a drain on society for many years to come, both costing money and contributing to environmental degradation.

Fortunately, the idea of green buildings in Asia has, in the past decade, moved from an esoteric concept to an increasingly mainstream practice.

Green building councils have been established in most East Asian countries, and stricter standards are being put in place, even if the pace of change—and of implementation—often seems maddeningly slow. Government-mandated building standards are the fastest and best way to effect fundamental change. Indeed, many of the best companies go far beyond mandatory standards and don't even talk about green design anymore. For example, Singapore-based Singbridge simply incorporates more environmentally conscious design principles into all of its massive new city projects as a matter of course.

Urban density makes energy efficiency in transportation possible. Transportation is about far more than cities, but it will be significant progress if more energy-efficient vehicles win broad acceptance in cities. Singapore, which combines a time-of-day road-pricing system and high registration fees for private cars, is a global pioneer in traffic management. But no other cities in Asia have yet been willing to follow Singapore in taking the politically difficult step of making private autos pay their own way.

Public transportation in Asia is good, but there is still room to extend existing networks as well as to build new systems for cities like Jakarta. Asian rail and subway systems are the world's largest and fastest; Hong Kong's MTR is among the best subways in the world, measured by both environmental and operating efficiencies. Public transportation will never replace cars, and every Asian city struggles to manage private car usage because of the congestion and the pollution as well as the greenhouse gas emissions that private automobiles create. Electric cars have struggled to win widespread acceptance among consumers. However, if electric vehicles ever do gain a mass market, it is likely to be in cities. Because of their relatively short travel range between chargings, battery-powered cars are particularly well suited to the shorter distances that city dwellers typically drive.

A recurrent theme in the following three chapters is how necessity drives change. Small countries, particularly Singapore, have been among the most forward-thinking. Countries that are severely resource constrained, such as Japan and South Korea, have acted more quickly than larger, more resource-rich countries. Now, with resource constraints pressing on even the largest countries, China and, to a lesser extent, India are taking the first small steps toward sustainability in the world of cities, buildings, and transportation.

In the next chapter, we'll look at how the challenge of sustainability is being met by various cities—both those that currently exist and are trying to become more sustainable and the new "eco-cities" that are being built from the ground up with sustainability as a founding principle. The following chapter is devoted specifically to buildings, and the final chapter in this part looks at vehicles—especially the increasing number of cars—that move people around. Although these chapters sketch out the problems, they are, above all, about opportunities, opportunities that are being seized as a way of making a break from business as usual when it comes to urban spaces, buildings, and transportation.

3

Cities in a Garden

> God the first garden made, and the first city Cain.
>
> —ABRAHAM COWLEY

At the time of its independence in 1965, Singapore was, in the words of founding Prime Minister Lee Kuan Yew, a place where "litter and dirt, the stench of rotting food and the clutter and obstructions turned many parts of the city into slums." He set out to mold the country into a clean, orderly, and environmentally friendly city as part of a broader campaign to instill civic order. "After independence, I searched for some dramatic way to distinguish ourselves from other Third World countries," he wrote thirty-five years later. "I settled for a clean and green Singapore." He was extraordinarily successful. Today, the country has explicitly developed the notion of the city in a garden, turning on its head Abraham Cowley's notion that the city is a creation of the violent Cain.[1]

Fittingly enough, two of Singapore's unique attractions are the Botanic Garden, founded in the mid-nineteenth century and long one of the world's most renowned, and the newly built Gardens by the Bay, a multimillion-dollar indoor complex opened on reclaimed land near the city center. "I have been stunned and incredulous only once, down by the harbor-bay in Singapore," wrote *Financial Times*' gardening columnist and Oxford classicist Robin Lane Fox in 2013, shortly after the gardens opened.[2] On a sprawling 133-acre site with more than a million plants, the country has created one of the most innovative botanical gardens anywhere, a jewel in the crown that it wears as one of the world's greenest cities.

Gardens by the Bay, Singapore's $800 million botanical garden and public park, which opened in 2012. Photo credit: David McIntyre

Singapore's commitment to environmental sustainability goes far beyond these two botanical gardens, however impressive they are. Singapore, a tiny country about the size of New York City, increased its greenery cover from 36 percent in 1986 to 46 percent in 2007—despite the population growing from 2.7 million to 4.6 million during that time.[3] The government mandated green building standards and has made them progressively tighter over the past decade. The country's electronic road-pricing system, part of a series of policies to limit cars that also includes competitive auctions for the right to own a car, was one of the world's earliest and most successful.[4] The city itself has been planned comprehensively since independence, with an increased focus in recent years on environmental sustainability. A city-state can act more quickly than a large country, of course, but Singapore's actions have emerged as a model for what can be done throughout Asia, even in China or India.

Singapore has long put issues of sustainability at the core of its national identity, thanks largely to Lee's conviction that the country faced existential threats because of its isolation from its natural hinterland, Malaysia.

After separating from Malaysia, the country became "a heart without a body," Lee wrote in his memoirs. "Seventy-five percent of our population of two million were Chinese, a tiny minority in an archipelago of 30,000 islands inhabited by more than 100 million Malay or Indonesian Muslims. We were a Chinese island in a Malay sea. How could we survive in such a hostile environment?"[5]

Singapore's most glaring strategic vulnerability is its lack of fresh water. Part of the answer to Lee's question came in the form of a thoroughgoing attempt to reduce the country's dependence on Malaysia for water (an effort that is described at length in chapter 6). For many years, it was heavily reliant on Malaysia, but now recycled water—sewage water, much of it—and desalination can supply more than half the city's water needs.

Because its very survival as a nation depended on it, no country in Asia and few in the world have embraced the concept of sustainability as thoroughly as Singapore. It is singular in its national commitment to addressing sustainability issues and to building manufacturing and services businesses, backed by government-nurtured research and development capacity, and this has enabled it to take advantage of green business opportunities. Thanks to this, the country has emerged as a global center of expertise in a variety of environmental services areas, from water to city planning. Long-established Singaporean companies such as Keppel and SembCorp have shifted away from their core engineering expertise to develop environmental-services-based businesses, while newer firms such as Hyflux have emerged as significant players in the area of water treatment and desalination. Now, the country is exporting its urban design and environmental expertise. Companies in Japan, Korea, and Taiwan have all used its technical and engineering expertise to take advantage of Asia's need to support more people at a higher standard of living, while using fewer resources. Singbridge, a subsidiary of the country's $170 billion Temasek sovereign wealth fund, is building eco-cities in China that we'll visit later in this chapter.

Urban developments focused on sustainability couldn't be more needed by the region. Asia is a place of more cities, and more densely populated cities, than anywhere in the world. Beijing, Hong Kong, Jakarta, Singapore, Seoul, and Manila—these six East Asian metropolitan areas together hold more than 100 million people. Greater Tokyo adds another 37 million inhabitants. Add a few other Japanese metropolitan areas (Osaka and Nagoya) and three of China's largest conurbations (Shanghai, Guangzhou, and Shenzhen), and the total grows to 200 million.

A dozen cities totaling 200 million people: this adds up to a combined population greater than all but four countries in the world. Consider that it was only in 1968 that the total U.S. population passed 200 million—or that the entire world's population was 200 million when Augustus founded the Roman Empire 2,000 years ago.[6] But it doesn't stop there. The five largest metropolitan areas in India, Pakistan, and Bangladesh add another 83 million people—a population larger than Germany's.

And it's not just that Asia's cities are large—they are growing, both in size and in number. McKinsey & Co., which has done an impressive and detailed series of studies on the world's top 2,000 cities, also looked at which metropolitan centers will become megacities of more than ten million by 2025. Only one, Chicago, is in a developed country. Ten of the remaining twelve cities are in Asia—cities that for the most part today are little known to the rest of the world. They include Chengdu, Dongguan, Guangzhou, Hangzhou, Shenzhen, Tianjin, and Wuhan in China; others are Jakarta (Indonesia), Lahore (Pakistan), and Chennai (India). Two African cities, Lagos (Nigeria) and Kinshasa (Congo), round out the list.[7]

Nobel Prize–winning Swedish economist Gunnar Myrdal teased out the numbers on population and economic growth in more than a dozen countries almost five decades ago and concluded that Asia would forever struggle to escape a vicious cycle of poverty as economic growth battled to outpace surging populations. Growth throughout much of Asia since then has been far higher than Myrdal had any reason to think possible—or indeed than the world had ever seen previously. McKinsey estimates that 225 cities in the China region (including Hong Kong and Taiwan) will by themselves be responsible for 30 percent of global growth from 2007 to 2025. And yet Myrdal's magisterial *Asian Drama: An Inquiry into the Poverty of Nations* contains a message that remains important today. How do we provide places for the Asian population to live, work, and play as Asia gets richer and more populous at the same time as air, water, and living space become ever scarcer? Can we invert Abraham Cowley's notion of the garden as paradise and the city as chaos and make our cities into earthly paradises?

This is a lofty goal, certainly, but also an increasingly urgent one. Asia's cities need to develop in an environmentally sound way, or much of the economic and social progress of recent decades is in danger of being eroded. Asia's cities are generally growing faster than their country averages, and these countries are mostly growing faster than global averages. That adds up to turbocharged growth—and with this population growth come new buildings, new roads, new mass transit systems, and new cities.

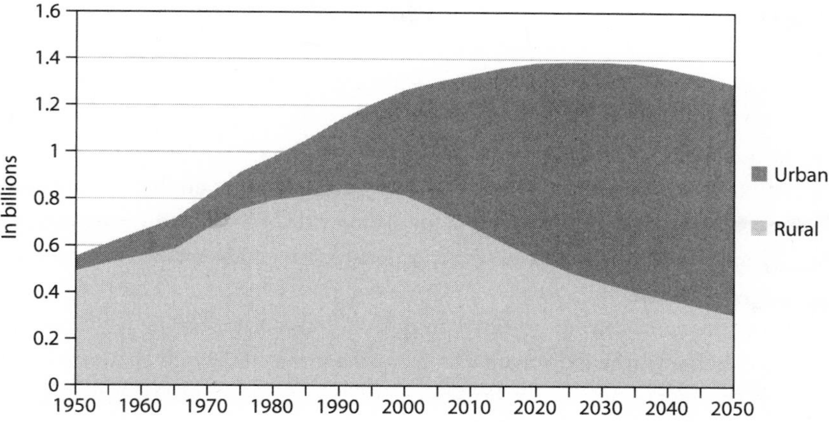

China's Urbanization
China's urbanization, the largest ever, will put new pressure on resources. Source: United Nations Population Division, World Urbanization Prospects

China is building the equivalent of a new Chicago every year, notes McKinsey director, Jonathan Woetzel, who estimates that the country's urban population will swell from its current 700 million to about one billion by 2025. But at what environmental and social costs? "Planning is done in a cookie-cutter approach with local authorities copying one another in a lemming-like fashion, reflecting the shortage of planning skills and the pressure on cities to cope with migration as well as the state's easy access to capital," notes Woetzel. "This results in overcapacity and, often, inappropriate design that is essentially wasted. . . . Socially, cities are becoming incubators of unrest as slowing urban job creation and dependence on heavy industry for economic growth threaten to create a two-tier society."[8]

Woetzel suggests that China should concentrate less on building new, smaller cities and focus on eleven larger hub-and-spoke clusters, such as the greater Pearl River Delta and the Yangtze Delta cluster centered on Shanghai. In an optimistic version of this scenario, Woetzel estimates that 40 percent of the population could live in these agglomerations, which are far more environmentally efficient and economically productive than a pattern of many smaller, scattered cities.[9] Woetzel and his McKinsey colleagues think that there is no fixed limit beyond which cities cannot grow ly. "The only hurdle to the growth of urban centers is an inability e with, and manage, their expansion. Large urban centers are complex and demanding environments that require a long planning horizon and extraordinary managerial skills. Yet many city governments are not prepared to cope with the speed at which their populations are growing. Without skillful planning and management, cities run the risk of diseconomies—such as congestion and pollution—starting to outweigh scale benefits, leading to a deteriorating quality of life and a loss of economic dynamism."[10]

The pace of change is hard to grasp, even for people who live in China. That country is responsible for more than half of all the construction in the world and a similar share of skyscrapers and tall buildings.[11] Building sites are not the only new construction. Woetzel notes that China could build as many as 170 mass transit systems over the next twenty years—45 are now under construction—compared with 50 existing in all of Europe now.[12] Indeed, virtually every East Asian capital city already has a significant mass transit rail system or plans one. "Across cities, the advent of national high-speed rail will link together the megacities even more, resulting in the development of a uniquely Chinese megacity culture and the rapid dissemination of best practices in urban development and management," says Woetzel.

However, at present, China's cities are decidedly neither efficient nor environmentally benign. Its buildings last only about twenty years versus a global average of forty, notes Woetzel, worsening the strain on resources needed to build them. Per capita CO_2 emissions in Beijing, Tianjin, and Shanghai are higher than in most global cities, including New York, Tokyo, London, and Hong Kong—and even, somewhat surprisingly, in Los Angeles. A combination of dirty power plants, continued heavy industry (especially cement and petrochemical plants), and transport all add up to a dismal situation.

Big cities present problems beyond just unsustainable growth and infrastructure. In the nineteenth and much of twentieth centuries, Asia's cities were the breeding places for diseases from cholera to the infamous 1918 Spanish flu. More recently, an outbreak of plague in Surat, India (in 1994), the occurrence of SARs in Guangdong (in 2003), and continuing concern over avian flu and swine flu throughout the region have served as reminders that the battle against disease is never-ending, especially in Asia's crowded cities, where population densities and the proximity of agricultural areas have contributed to the development of new human maladies.

Lee Kuan Yew focused on environmental cleanups in part because of his concern with diseases, such as tuberculosis.[13]

Despite the challenges that cities present, they are the only answer for an Asia whose 4 billion-plus people today likely will reach some 5.2 billion around the middle of the century. The question is how to get the most benefits from cities.

For a start, modern sanitation systems and efficient public transport have helped make cities prosperous and often even pleasant places. People are healthier and wealthier in cities. City dwellers generally live longer than their country cousins. There are more doctors and nurses and more hospital and clinics, so health care is better. People in cities are less likely to die of accidents, especially automobile accidents.[14]

Cities also have the potential to be more economically efficient and more resource-efficient. McKinsey estimates that it costs only half as much to deliver piped water in cities as elsewhere. Studies show that a doubling of density results in a 30 percent reduction in energy use per capita—although people in cities do consume more than their rural cousins. Many big infrastructure projects, such as airports and mass transit systems, need a critical mass of people to be economical.[15]

So the question becomes this: Will some combination of pollution, congestion, expensive housing, and crumbling and overloaded infrastructure stifle growth—or will the citizens of a richer Asia demand better cities? Urban economics expert Edward Glaeser warns that there are three cardinal principles for successful cities: governments should make it easy to build housing, see that good infrastructure is constructed, and ensure personal and public safety. Governments need to make it fairly easy to build, and (perhaps counterintuitively) to build densely, in order to allow populations to grow and to keep real estate prices from being a brake on economic activity. Heritage must be balanced with the need for renewal and the need for density.[16]

Asia's cities are still in the process of being built, their final forms unshaped, allowing for choices of a sort that more mature economies no longer enjoy. And some of them, at least, seem eager to forge a new urban model.

Perhaps the most expensive urban project per resident is Masdar City, in Abu Dhabi.[17] The small desert city, designed by star British architect Norman Foster's eponymous Foster + Partners and intended to be home to some 40,000 residents by 2025, is being built at an estimated cost of $18 to $19 billion over twenty years. This project is being funded by Abu Dhabi, the largest and wealthiest state in the United Arab Emirates; it claims to hold up to 8 percent of the world's proven oil reserves. That a petro-state is building an eco-city in the desert might seem ironic, but it is in fact an example of the long-term thinking that small countries ignore at their peril. Abu Dhabi is trying to wean its economy from its overwhelming dependence on oil revenues and to build up a new area of economic expertise in preparation for the day when fossil fuels no longer power the world.

Masdar residents will live in a pedestrian-oriented city built using sustainable building materials and powered by geothermal and solar energy transmitted over a smart grid. The enclave's buildings will use less than half the electricity and water of conventional buildings—the latter is especially important, as Abu Dhabi faces severe water shortages. To get around, the city's residents and visitors alike will be able to drive Mitsubishi electric cars or even an experimental driverless vehicle.

Masdar has attracted significant interest from major global companies. Siemens, which has smart-grid and intelligent building projects under way in the city, has already built its Middle East headquarters there. Mitsubishi Heavy Industries has teamed up with the city to study electric vehicles' performance in the desert climate. The city has enlisted Broad and Schneider Electric, among others, to develop a more efficient solar-powered air conditioner, and General Electric is working with it on smart appliances. One of many solar projects has the Masdar Institute of Technology collaborating with the Tokyo Institute for Technology and Japan's Cosmo Oil Company.

Masdar may be the most expensive project, but cities in many countries are reaching out to some of the world's leading experts on sustainability issues to pursue projects with a similar aim. The U.S. architecture firm Skidmore, Owings & Merrill (SOM) is doing a master plan for Beijing.[18] The city of Changsha is developing a new lakeside district for 180,000 people following a master plan by Kohn Pedersen Fox, which in turn draws on Gale International's experience in Songdo, Korea (a project discussed at greater length below). Throughout the country, China is piloting a number of low-carbon programs, such as the Tianjin Eco-city discussed below. In the following pages, we will look at how some of the most innovative and promising cities and corporations in Asia are tackling the question of green growth.

~

If you can build an eco-city on the toxic outskirts of Tianjin, you can build one anywhere: that was the thinking behind a unique Sino-Singapore joint venture in choosing a site for a new green city that, even in China's polluted and degraded landscape, stood out for its sheer awfulness.[19]

Large parts of the area were underwater. Untreated sewage had been dumped there for decades. Salt drying took place on the edge of brackish ponds. Even grass would not grow without help. Drinking water from this blighted landscape was a distant dream—even to touch it was a human health risk.

"They insisted that whatever [site] we chose had to have very difficult conditions environmentally so that when we made it a success they would be able to replicate it elsewhere," said Lim Chee Onn, a Singaporean who, as head of project developer Singbridge, played a key role in getting the project started. The Chinese government, one of the funders of the project, laid down two conditions: the venture could not take over agricultural land, and there must be an inadequate water supply. "This is like boxing with your hands tied behind your back," said Lim. "My conditions were that it had to be near economic activities. To implement environmental improvements you need people. . . . Second, I wanted cheap land. I didn't want to be holding expensive land. If you have a dip [in the economy] you will be growing very expensive grass." This was the challenge that Singapore's Singbridge and its Chinese partners faced as they plotted how to develop an area of thirty square kilometers, a little more than half the size of Manhattan, with an initial development budget of ¥50 billion, or a little more than $8 billion.

Asia is unique in its development of these high-profile new cities—so-called greenfield projects—which are being built according to detailed urban plans that incorporate a variety of good environmental practices. There are certainly examples in history of cities being massively rebuilt; think of the rebuilding of Paris in the second half of the nineteenth century under Emperor Napoleon III and Georges-Eugène Haussmann. But the idea of building cities from the ground up, of taking more or less unused land and creating a city on it the way a painter creates a painting on a blank canvas, is a more recent conceit. The Brazilian capital of Brasilia, founded in 1960, is a famous early example.

The Singaporeans and the Chinese are building the Tianjin Eco-city as a demonstration project on the Bohai Gulf, 150 kilometers southeast of central Beijing. Tianjin, like Beijing, Shanghai, and Chongqing, is a self-governing municipality with provincial-level status. In fact, with a population of more than thirteen million and an area three times as large as Rhode Island, it is more like a small province than a city. Reporting directly to the central government, Tianjin enjoys special treatment. It also benefits from good infrastructure links. A high-speed expressway connects it to the capital. An even faster alternative is a high-speed train, running at over 300 kilometers per hour, that makes the trip from Beijing to downtown Tianjin in forty-five minutes.

On a site where the water was unfit for human touch, the Tianjin Eco-city promises to supply water that its residents can drink from the tap. This would be an audacious goal anywhere in China, a country where those who can afford to buy bottled water do so without fail, and those who can't afford bottled water always boil theirs first. Underscoring the difficulty of ensuring good water quality, the nearby Holiday Inn Binhai Tianjin displays a prominent sign warning guests not to drink water from the tap in their room: "Water Is Not Drinkable Before Boiled."

Water is just one challenge that the builders of the Tianjin Eco-city face. Every building must adhere to at least minimal green building standards. To encourage better buildings, the Singapore and Chinese governments in 2013 agreed to fund up to $12 million in research over the next three years on green buildings, using the eco-city as a laboratory. Another hurdle is that the residents of the buildings must follow sustainability guidelines to help the city meet its eco-goals. This is a major challenge in and of itself, with apartment owners who often are more eager to show their newfound affluence with conspicuous consumption than to pursue an eco-friendly lifestyle. The city is expected to be home to 350,000 people by the time it is built out in 2020; as a living city, will residents embrace concepts of environmental sustainability in their daily lives?

Former Singbridge CEO Ko Kheng Hwa says that setting hard targets—key performance indicators (KPIs)—will ensure adherence to the notion of sustainability. The eco-city, backed by the two countries' respective governments, has approved twenty-six of these indicators, across a wide range of fields, from the percentage of trips that are green to the percentages of recycled waste and recycled water. "Setting these KPIs is part of the holistic approach in the planning and execution of a greenfield city," says Ko.[20]

Some of the targets, especially for individual behavior, are quite stringent. Daily water consumption should not exceed 120 liters per person by 2020—much lower than Singapore's 153 liters per capita in 2011 and still well below its 2030 daily target of 140 liters per head.[21] Domestic waste should not exceed 0.8 kilograms per person (a little less than two pounds) per day, starting with the opening of the project in 2013. From the beginning, all buildings should have barrier-free access, and all residential buildings should be within 500 meters of a free recreational and sports area; there will be public green space of at least 12 square meters per person. Renewable energy should be 20 percent of the total by 2020; that same year, 50 percent of the water should come from nontraditional sources, such as recycling or desalination. To try to ensure that the eco-city is economically self-sustaining, the aim is that by 2020 half of those who live in the city will work in the city. "To the extent we can create economic growth that enables the government to collect more revenue thereby enabling them to invest in green infrastructure, that is another form of sustainability," says Ko. "If you have a green city or an eco-city and you are unable to provide the economic activities, there comes a time when you ask what people are going to do."

Singapore has a track record of doing a lot with few resources. Many Chinese officials trained there as a result of Deng Xiaoping's admiration for its ability to survive in a situation with few resources in an insecure neighborhood. Singapore "was a country with nothing—we had to import water," remembers Lim Chee Onn. The country first got into offshore city-building in the mid-1990s when Lee Kuan Yew, convinced that it was not enough to train Chinese officials in Singapore because there would never be enough of them to make up a critical mass able to push through change, sought a large-scale project in China. As a result, Singapore's Keppel Corp. built a township and industrial park in the historic Chinese city of Suzhou. Keppel, one of Singapore's largest employers, was a natural choice; it started as a ship repair business but parlayed its maritime and engineering expertise into, among other things, energy-efficient designs in its property sector. Despite Lee negotiating personally with his Chinese counterparts, there were numerous conflicts, culminating in Suzhou authorities setting up a rival industrial park. Nonetheless, Lim believes the Singaporean emphasis on high design and construction standards as well as a high degree of transparency paid dividends. "Suzhou was very useful from a political point of view," he says. "It created a platform without parallel between Singapore and China."[22]

Fifteen years after construction on Suzhou started, talks about a new joint project began. Former Singaporean Prime Minister Goh Chok Tong mentioned the idea to Chinese Premier Wen Jiabao at a meeting in Beijing in April 2007. Goh also called Lim Chee Onn, who was the head of Keppel at the time and had been a key player in the development of the Suzhou project. The result was the Tianjin Eco-city, a larger-scale project that was designed to be an explicitly environmental city showcasing an integrated building and transportation network. "China was very keen to show the world about the seriousness of its efforts to deal with its pollution problems," remembers Lim, who left Keppel to become Singbridge's first head after it was founded in 2007. Underscoring the high-level commitment to the project, the November 2007 framework agreement was signed by Wen and by Singaporean Prime Minister Lee Hsien Loong.

Singapore's city-building activities have paralleled China's changing priorities. When Suzhou was built in the mid-1990s, the priority was industry. Today, the focus is on sustainability, and in Tianjin, the goal is to build an ecologically viable city on an environmentally challenging site, one with a shortage of water and severely degraded soil. Looking forward, Singbridge has a large-scale project in Jilin that is intended to upgrade a county-sized rural area and bring its farmers and food processors into the world of more modern and hygienic technology. At the same time, it reflects Singapore's newfound concern with food security.[23]

"We have excellent prospects," said Ko Kheng Hwa, "because we are riding on three global trends. One is massive urbanization. The second part of it is a much stronger desire for environmental enhancement—in some parts driven by governments and in some parts the citizenry drive the governments. Third, in Asian countries there is this need for economic development. Our projects need to be scalable, replicable, and commercially viable."

Singbridge's Lim Chee Onn was adamant that he would not build an eco-city in an isolated area because without economic activities the idea of a sustainable city is meaningless. "What is green? You can have a zero carbon footprint—but is that practical?" asks Lim. "You cannot be doing zero carbon and everyone living in tents without lights. What is your ultimate objective? It must be improving the standard of living."

~

Korea's Songdo new city is another example of a greenfield site that is designed to be an environmentally friendly, hip, modern, and fast-wired city.[24] Songdo is being built on a vast twenty-one-square-mile expanse of reclaimed land on Korea's coast in Incheon, forty miles west of Seoul, where General Douglas MacArthur led one of the most famous landing invasions in military history during the Korean War. Within the city is an international business district whose 1,500 acres of land are connected by an eight-mile bridge to Incheon International Airport. The airport is among

The Songdo International Business District is a smart city built from scratch on 1,500 acres of reclaimed land along the waterfront of Incheon, South Korea. More than 40 percent of its area is reserved for green space, and all of its major buildings meet or exceed LEED expectations. Photo credit: Gale International

the dozen busiest in the world, and this proximity makes the possibility of Songdo's success more plausible.

Songdo's credentials were burnished by Korean rapper Psy, who shot parts of his "Gangnam Style" hit there (for those who know the video, it's the garage, the elevator, and the exploding field). Psy subsequently made a parody video, "Songdo Style," that was part of a successful 2012 bid to convince the United Nations Green Climate Fund to set up its global secretariat in the city. Green Climate Fund staff began arriving there in 2014, and they were expected to be followed by employees from the World Bank, one of the first times that the bank has set up outside of a capital city. Songdo backers hope that these companies will act as anchors in their bid to lure more tenants.

"The Green Climate Fund coming in put us on the global map for sustainability," says Scott Summers, vice president of marketing and investment at Songdo developer Gale International. Summers notes that few other Asian cities can boast of schools, housing, and offices within a fifteen-minute walk of one another. Indeed, designers worked hard to ensure that most daily activities would be within this fifteen-minute radius, the limit before people typically get in their cars.

A key element in Songdo's attempt to differentiate itself from Seoul and from other competing cities is its environmental credentials. Now home to 30,000 people, the city claims that it will emit only one third the greenhouse gases per person as traditional cities and will use significantly less water and electricity.

All Songdo buildings must be built to levels set by the U.S. Green Building Council's Leadership in Energy Efficiency and Design (LEED) standards, the only exception being the first building, which is not LEED certified. All this makes it the world's largest LEED-certified city. It is home to Korea's first LEED-certified hotel and first LEED-certified school and to Asia's first LEED-certified convention center.

The municipal government has built cogeneration facilities for the entire twenty-one-square-mile city; these facilities use waste heat to produce hot water, which is used in traditional Korean underfloor (*ondol*) heating. Waste is whisked by pneumatic tubes from apartment buildings to imaginatively landscaped waste transfer stations on the edge of the district, eliminating the need for garbage trucks to come into the city center. Water is captured, treated, and recycled for irrigation and street washing. "Korea has a water issue and water conservation is a big component of Songdo," says Summers.

Songdo's environmental credentials are only one part of an attempt to create an appealing, commercially viable living environment. Some 40 percent of the project, including a 100-acre Central Park, is set aside as parkland and waterways, in accordance with a master plan laid out by architects Kohn Pedersen Fox. A new arts center is being built and is accessible from the city center through Central Park on foot or by water taxi. An Arnold Palmer–designed golf course will host the prestigious Presidents Cup in 2015, the first time the event will be played in Asia. An adjacent area of Songdo hosts the Korean outposts of the State University of New York (SUNY, Korea), George Mason University, and other international universities.

Songdo is a commercial project, one that enjoys powerful backing. South Korean steelmaker Posco—formerly state-owned and still enjoying powerful political connections—holds 30 percent of the project, with New York–based property developer Gale International leading a consortium that holds the other 70 percent.

When Songdo is fully built out, with 2017 as the target date, the city is designed to have 300,000 commuters and 65,000 residents. The numbers only begin to hint at the enormity of this project. Upon completion, the complex is expected to have 50,000 apartments, fifty million square feet of office space, and ten million square feet of retail space. This is an enormous amount of new space for any city, let alone a new one, and even some of those who have invested in the project, or bought space for their companies, expect that it will be some time before the city is able to prove its economic viability. New York City struggled for more than a decade to fill the original World Trade Center towers after they were completed in the early 1970s—and they had less than one tenth the floor space of Songdo's planned new office and retail space.

Metropolitan Seoul already has about twenty-five million people, or about half of the country's population. Developing Songdo as a new center within a vast metropolitan region, especially given its proximity to the country's most important airport, means that South Korea's largest satellite city would seem to have a good chance of success. The environmental and design principles underlying Songdo's development were seen as a way of making the project more commercially viable, thus increasing its chances of success.

~

A modest, four-story, rented building tucked away among the tangled streets of northern Mumbai's sprawl might seem like an odd stop for French President François Hollande to make on his first state visit to India. Yet Hollande thought the site important enough to make time, during a brief five-hour stopover in India's commercial and financial capital, to thread his way through the skein of narrow, erratically paved streets to visit the unusual building—notable because its flat concrete roof is capped with a peasant's mud hut.[25]

This building houses the development laboratory set up by France's Lafarge, the world's largest cement company in terms of sales. Hollande made this out-of-the-way pilgrimage because he knows the cement and concrete that French companies like Lafarge supply will become increasingly important as Asian cities push to build the new airports, bridges, office towers, hotels, and apartment buildings that their fast growth requires. Tellingly, the secret talks that led to Lafarge's 2014 merger agreement with longtime Swiss rival Holcim were code-named "Cities."[26] It's not just urban areas that Lafarge will affect; the peasant's hut on the top of the building demonstrated how low-cost Lafarge materials could extend the life span of traditional huts and improve the overall livability of India's ubiquitous hut villages.

This low-key compound is just one of a growing number of development laboratories Lafarge has set up in order to better sell its products in fast-growing local markets. Europe's construction market is slumbering, and construction in rich countries does little more than mirror overall economic growth. Even before announcement of the merger with Holcim, Lafarge was selling off Old World assets to cut debt and focus on the emerging markets of Asia and Africa, a strategy that will be accelerated as part of the planned merger.

Lafarge's growth strategy is tied to those parts of the world where new building is accelerating—above all, India and China. Paralleling the strategy at the Mumbai lab, Lafarge recently opened a similar center in Chongqing, China, where the company is working with municipal authorities to build large tracts of low-income housing. With worldwide capacity to make almost 225 million tons of cement a year, Lafarge's production capacity is larger than that of most countries, excepting China (1,452 million tons of capacity) and India (301 million tons). Indeed, Paris-based Lafarge has twice the capacity of the United States (114 million tons of capacity).[27]

"There are two areas where urbanization [most] matters," says Lafarge Chairman and CEO Bruno Lafont, "and that is Asia and Africa." Singling out China, India, Indonesia, Vietnam, and Bangladesh in Asia, Lafont adds that "there are huge issues regarding housing, infrastructure and the migration of people and that is where we will find growth. Our goal is to contribute to this challenge positively, to contribute to building better cities."

Cement making is not a pretty process. It requires vast quantities of energy to quarry limestone, which is then mixed with coal and cooked to 2,700 degrees Fahrenheit in giant kilns—cylindrical tubes wide enough to drive a car in and with a horizontal length equivalent to the height of a multistory building. The process produces a sandy substance with a very particular chemical composition that, when mixed with water, makes cement—a material with extraordinary strength and plasticity that can be sculpted into almost any shape.

The environmental challenges facing the industry, considering how energy- and material-intense the process is, are daunting to say the least. Today, the cement industry alone accounts for 5 percent of total greenhouse gas emissions.

Given this, is the very idea of an environmentally benign cement company an illusion? Certainly, the cement industry by definition cannot be environmentally neutral unless it uses recycled raw materials—rather than quarrying limestone and other ingredients—and generates its heat using alternative sources. Lafont is outspoken in his belief that the industry must do better environmentally and has committed Lafarge to cutting the amount of carbon emitted per ton of cement by one third between 1990 and 2020. By the end of 2012, the company had already cut its so-called carbon intensity (the amount of carbon emitted per ton of production) by 24 percent. Lafont says that the company is "more than on track" to reach the 33 percent goal by the end of the decade.

Lafarge has achieved this increased carbon efficiency in a variety of ways. It has increased the energy efficiency of its kilns. Rather than relying primarily on limestone in the process of manufacturing cement, it increasingly uses industrial waste such as fly ash, the by-product from coal that is burned to produce electricity. This is a tidy solution to the problem power plants face in figuring out where to dump their mountains of fly ash. Lafarge also uses more nonfossil fuels than it once did, such as waste from industrial plants or even household waste.[28]

"Energy efficiency is a highly theoretical concept driven by multinational companies with carbon targets," says Lionel Bourbon, who runs the Mumbai lab. What really counts for construction companies and the people who buy their buildings is time and money, he says, pointing to an aluminum coffer framework that reduces the amount of steel used in construction as an example of the type of product that research-focused companies like Lafarge can develop. This framework, designed for six-story mass housing blocks, in turn supports a lightweight concrete with improved insulating properties. The result is better energy efficiency at what Lafarge says is a 9 percent cost reduction.

These sorts of innovations mean more comfortable, more energy-efficient, and more economical housing for Asia's ever-wealthier and more numerous citizens. They mean better office buildings and shops. "There is no urbanization without cement and without concrete," says Lafont, who previously chaired the Cement Sustainability Initiative of the World Council on Sustainable Development.

The challenge, however, is to dramatically improve the industry's environmental record. That's especially true in developing countries, countries that account for 80 percent of cement-related greenhouse gas emissions. China is responsible for more than half of cement production and likely at least that proportion of emissions. (The data on China's cement-related emissions are incomplete because less than 20 percent of production is represented in the Cement Sustainability Initiative, the members of which are required to report their numbers and encouraged to have outside verification.)

India is the world's second-largest cement market after China, with an annual production of 300 million tons in 2010 expected to almost double to 550 million tons by 2020.[29] Housing accounts for two thirds of the demand. Indeed, homeowners in India will require ten million new dwellings a year by 2030. This annual production is equivalent to about three times the number of existing units in New York City.[30]

In markets like India, Lafarge competes against what it says is low-quality cement and aggregate that is mixed on site. The erratic quality of this on-site cement leads builders and buyers to overengineer the buildings and to use more cement than they would otherwise need to. Lafarge's obviously self-interested pitch aside, a greener world will be one in which materials will be used sparingly, not excessively because quality is suspect. In any event, Lafont is optimistic about the company's future in countries like India. It is now working on a large as well as a very small scale, delivering cement a gallon at a time, by motorcycle, using a just-in-time system modeled on pizza deliveries.

Lafarge is by no means alone in its desire to reduce both the costs and the environmental impact of cement manufacturing. Among Asia-based companies, Siam Cement is one of the most notable sustainability leaders. It was founded by the Thai monarchy in 1913 and, a century later, has grown into one of Southeast Asia's largest companies. Now known as SCG, the group's activities include chemicals, which today make up half the business; building materials; paper; and, of course, cement, with 2013 annual sales of $13 billion.[31] Its SCG Building Materials & Fixtures unit was named the best in its sector in 2011, 2012, and 2013 by the Dow Jones Sustainability Index. Like Lafarge, SCG is a member of the Cement Sustainability Initiative.

In 1995, SCG set up a sustainable development committee, and in 2008, it laid out hard targets for its various business units, promising to reduce the intensity of greenhouse gas emissions by 10 percent by 2020, using 2007 as a base. It reached its goal of essentially eliminating industrial waste going to landfills by 2012, largely by burning everything from tires to rice husks for use in its industrial processes.[32] The company's goal of being recognized as a global environmental leader in its various businesses is well within reach, yet SCG's expansion exposes the dilemma facing fast-growing Asian companies. Production has increased so dramatically that even impressive efficiency gains in some areas, such as emissions from its cement plants, are outweighed by larger volume. "An absolute reduction in emissions is not possible for a developing country," says company executive Cholathorn Dumrongsak.

For, the reality remains that the cement industry is a major contributor both to air pollution and to greenhouse gas emissions. Yet life as we know it for seven billion people—life with high-rise buildings, bridges, and roads—would not exist without cement and its final form, concrete. Can the cement industry support global growth and the desire for better places to live and work in a less environmentally destructive fashion? Whatever progress companies like Lafarge and SCG make is unlikely to be meaningful in a global context without significant changes in India and China, the two big producers.

China's Institute of Public and Environmental Affairs (IPE), one of the country's best-known environmental groups, teamed up with other domestic nongovernmental organizations to try to ascertain the environmental policies of seventeen publicly listed cement companies in China, sending out letters requesting such policies. The only company that indicated a willingness to follow up was Lafarge. Replies from the other sixteen ranged from "If you have not received a reply to this letter it is probably because the company felt that the contents of the letter were of no interest" to "If we feel it's necessary we will follow up and contact you."[33]

With urbanization growing and Asian cities experiencing a disproportionate share of growth, it is imperative that most companies in the cement industry begin adopting policies that will improve their carbon intensity and embracing the concept of sustainability.

~

Established cities, notably Singapore, have adopted the idea of sustainability and are already showing measurable results. There are the greenfield eco-cities, such as China's Tianjin and Korea's Songdo, that incorporate many promising elements of sustainable urban design as well as a variety of environmental technologies. But these greenfield cities are, by definition, untested, and their long-run environmental footprint, as well as their broader impact as demonstration projects intended to inspire the nation or even the region, remains in doubt. Moreover, these demonstration projects are small. Their commitment to sustainability must be matched by existing cities if there's to be any hope of large-scale change.

One of the cities to look to, besides Singapore, is Hong Kong. Christine Loh, who in 2012 was appointed Hong Kong's undersecretary for the environment, is outspoken in her assertion that Hong Kong and neighboring China will see dramatic environmental improvements in the next decade.[34]

In 1992, when Loh was a Hong Kong legislator, "the Hong Kong Harbor was effectively a public toilet," with most of Hong Kong Island's sewage dumped untreated into the water. Now, after two decades and billions of dollars in investment, the water is clean enough that in 2011 a cross-harbor swim, dormant since 1978, was revived—an outcome particularly exciting to Loh, who also is a founder and former head of the Society for Protection of the Harbor. In terms of environmental successes on land, Loh points to a number of green initiatives: the more than $4 billion that the Hong Kong government will spend on solid waste treatment over the next decade; the policies designed to encourage cycling and make the city more pedestrian friendly; and the growing number of green walls, sky gardens, and experiments in urban agriculture. According to Loh:

> We have put in place good sanitation standards and reasonable environmental standards. The next phase that we have to do is to create the best conditions within a high-rise highly dense environment for public health. We need to have a completely different way of dealing with waste. Waste and sanitation—we are going to go on a dramatic upgrade for our whole waste management system. My colleagues in the health area have made great advances in dealing with infectious diseases. We are spending a lot of money to clean up basic things like air and water.

Hong Kong is acting as a benchmark for many Chinese cities. China "very much would like us to take a leadership role," Loh says, particularly in acting on air and water issues in partnership with neighboring Guangdong Province, also on the Pearl River Delta. Outside of Singapore and Hong Kong, there are few cities that have adopted a broad-based approach to sustainability. Most of those that have are concentrated in Japan, although many cities in Taiwan and South Korea have made significant progress.

Kyoto stands as a good example of the sort of progress that can be made. One of the world's most captivating urban areas and Japan's longtime imperial capital, the city is today considered the apotheosis of Japanese temple and garden design. As one of the oldest and most charming cities in Asia, its charm convinced U.S. military planners to spare it from the bombing campaign against Japan during World War II. But by the 1960s and 1970s, the city's historical and architectural beauty was in danger. Traditional buildings were torn down to make way for undistinguished concrete blocks. Its streetcar trolleys were shut down and the track ripped up. Its ancient, storied river was canalized and cut off from the city. Though it had escaped bombing, the city was on the brink of becoming another dreary modern city.

Today, Kyoto and its surrounding urban area are home to world-class manufacturers, and the older parts of the city house some of the most exquisite gardens, temples, and houses in the world. The river has been cleaned up and its cherry tree–lined banks opened to joggers, cyclists, and fisherman. Subway service started in 1981, providing a mass transit alternative to private automobiles.

The city acts as a promising example of how Asian cities can foster growth without sacrificing their historical character and natural environment. Kyoto Mayor Daisaku Kadokawa—a former teacher who typically wears a formal kimono for meetings—speaks passionately about the delicate balance between industry and trade, about the time-worn temples that dot the city and the tourist trade that makes up its economic lifeblood.[35] It is this balance—between the ancient and the modern and between industrial growth and environmental preservation—that Asia's cities must strive for in order to see a future that is both prosperous and sustainable.

Cities with an environmental focus will become far more livable in the next decade. Places like Hong Kong, Kyoto, Singapore, Tokyo, Seoul, and Taipei will build on their wealth and expertise and on their citizens' demands to make their living areas more environmentally friendly. There are many other good examples of cities taking steps in the right direction. New cities, like Tianjin's Eco-city and South Korea's Songdo, will likely become more prominent. However, these showcase projects, important as they might be, are less significant than implementing increasingly higher efficiency standards for older cities. It is here that countries with the best-developed rules of law have an advantage. It will not be by top-down edicts that the battle for energy efficiency—and, ultimately, energy security and a more pleasant living and working environment—will be won. It will be by transparent and increasingly stringent building energy-efficiency regulations, combined with genuine support from ordinary users and cooperation from the business community. The question for Asia, and for the world, is how many less environmentally conscious cities will develop in the region—and how quickly.

4

Buildings for a Greener Asia

> You talk about sustainability and people's eyes glaze over. You talk about energy efficiency and risk management and people say "I've got it."
>
> —PHILIPPE LACAMP

Three decades ago, most urban Chinese walked up the stairs to their apartment in a low-rise building, usually about six stories high. At home, they needed electricity only for a few lights and a handful of appliances—a refrigerator, a fan, an iron, a radio, and, for the more affluent, a television, and perhaps a VCR and a rice cooker.

Today, most of China's 700 million city dwellers take an elevator to an apartment in a high-rise building, often twenty or more stories. If they are at all typical, their brightly lit apartment has air-conditioning in most rooms; a large-screen television; an assortment of mobile phones, tablets, or computers; and a variety of electric appliances, from microwaves to hair dryers. Likewise, Chinese offices three decades ago were dark and uncomfortable, with even window-mounted air conditioners a luxury and a shared fax machine and printer the most significant electrical devices. Now Chinese offices, too, are bright and air-conditioned, and they are equipped with the latest technology. There are 500 million more Chinese city dwellers than there were three decades ago, and they use electricity on a par with their counterparts in the West.

Shanghai's Pudong district, across the Huangpu River from the Bund, is one of the most striking examples of how Asia's buildings have changed in a generation. As recently as 1993, the eastern bank of the Huangpu was lined with low-rise walk-up apartment buildings. But since the bulbous, 1,535-foot-high Oriental Pearl observation and communications tower redrew the Shanghai skyline in 1994, a cluster of some of the world's tallest skyscrapers has been built in Pudong. The 88-story Jin Mao Tower, housing the Grand Hyatt Shanghai, was topped by its next-door neighbor, the 101-story Shanghai World Financial Center, in 2007. A new building, the 128-story Shanghai Tower, topped out in August 2013. These are among the tallest buildings not only in China but also in the entire world.

Although Shanghai is perhaps the most dramatic example of a city that has gone from low-rise and mid-rise buildings to a jungle of skyscrapers in less than a generation, most of Asia's major cities tell a similar story. Beijing, Taipei, Hong Kong, and numerous smaller cities would scarcely be recognizable to someone who had not visited since the 1970s. Cities in Asia have spread both up and out, with the elevator functioning as a sort of vertical highway. The elevator-as-highway is more than metaphorical: at forty miles per hour, the elevators in the Shanghai Tower move faster than most of the city's auto traffic.[1]

As of mid-2014, 9 of the world's 10 tallest existing buildings are in Asia. An astonishing 46 of the 50 tallest buildings under construction are in Asia (the four non-Asian buildings are all in New York City). When it comes to sheer numbers of skyscrapers, China dominates. As recently as 1990, China had only 5 buildings taller than 200 meters, or roughly forty-five stories tall. By the end of 2013, there were more than 250 buildings over this height.[2]

The sky literally seems to be the limit for Asia's tall buildings, as companies continue to spawn ever-grander projects. The Broad Group has claimed it will build a particularly ambitious one in Changsha, near the birthplace of Mao Zedong. The company has a vision of building what is expected to be, if it is ever built, the world's second-tallest building, the Sky City Tower. At 2,749 feet, the building would be over half a mile high, more than twice as tall as the Empire State Building. Mao, fond of grand schemes, almost certainly would have approved of this audacious project.

This combination of more city dwellers with more money is putting enormous strains on Asia's energy infrastructure. This is because in Asian countries, as in most of the world, the building sector is the single largest electricity consumer. As an extreme example, buildings in Hong Kong account for about 90 percent of total electricity consumption and at least 60 percent of CO_2 emissions.[3] Unfortunately, most buildings constructed in Asia since the 1970s are energy hogs, with their inefficient lighting, heating, and cooling systems making them the concrete equivalent of a gas-guzzling Hummer. (Most of these buildings are actually worse than Hummers, for buildings as a whole use far more energy than cars do. The world's buildings are, directly and indirectly, responsible for close to double the annual CO_2 emissions of the world's entire vehicle fleet.)

The energy used in operating buildings—heating, cooling, and lighting—alone accounts for more than one third of global energy use and a similar amount of total CO_2 emissions. When the energy used to produce and transport a building's concrete, steel, and glass is included, buildings account for over 40 percent of total energy consumption globally. Cement maker Lafarge estimates that over the life cycle of a building 10 percent of the total energy is used in construction, about 80 percent is used by the occupants during the life of the building, and the rest is consumed at the end of a building's life. In Asia, the dramatically shortened life spans of many buildings—fifteen to twenty years, typically, compared with thirty to fifty years in other, more developed countries—changes that equation, so that construction accounts for a far higher percentage of the building's energy over its life cycle. These badly built structures amplify environmental challenges.[4]

In the last chapter, we looked at the importance of sustainable cities for Asia's overall energy and environmental solution. In this chapter, we consider one of the most important elements in the quest for more environmentally friendly cities—more sustainable buildings.

~

There are enormous opportunities for energy efficiency in the building sector because green buildings can dramatically reduce energy use. The Natural Resources Defense Council calculates that in China green buildings can cut electricity use by 25 to 50 percent, water use by 40 percent, and solid waste by 70 percent, compared to a conventional building. Buildings account for more than half of all electricity consumption in China, so cutting electricity use by 25 or 50 percent would have a significant effect on electrical demand—and on the need to build power stations. The potential energy savings are enormous. If all of China's buildings were energy efficient, the annual savings would be enough to provide more than four months' lighting in the United States. Of course, this is a theoretical and ideal number, but it does give a sense of how much energy could be saved through a transition to more efficient buildings.[5]

Moreover, even ignoring the dramatic environmental benefits, it is cheaper to put money into more energy-efficient buildings than to build new coal-fired power plants to provide power for conventional, inefficient buildings. In China, fabricating new buildings to higher energy-efficiency standards is more cost-effective than retrofitting old ones. To be more exact, the cost of saving a megawatt of demand by constructing and operating a building more efficiently is one third less than that of adding another megawatt of electricity by building more generating capacity—and that is not accounting for the environmental benefit of burning less fossil fuel. Yet China currently is building the equivalent of two or more 500 MW power plants every week. With China responsible for more than half of all the floor space built annually in the world, more energy-efficient buildings are the most cost-effective way to address energy and environmental challenges. The same is true throughout developing Asia.[6]

"It's all part of a process of explaining to people how much they can save if they do things in a certain way," says Jamshyd Godrej, the chairman of Mumbai-based Godrej & Boyce Manufacturing and a global leader in sustainable building. "The economic argument generally works very well for most owners and tenants because heating and cooling generally are the highest costs for running a building."

The idea of green buildings has been around for decades. One of the pioneers was American physicist turned environmental crusader Amory Lovins, who built a demonstration house at a 7,100-foot elevation in the Colorado Rockies in the early 1980s. Lovins pioneered the concept of negawatts, the idea that it is often cheaper to save a megawatt of electrical demand by investing in efficiency than it is to build a megawatt of additional capacity. Lovins's insight was that what matters is not the amount of electricity produced but the end result: having comfortable offices and homes with good lighting and appliances that get enough electricity to do their job.

Asia has too few people like Lovins—people who are both scientifically knowledgeable and passionate advocates for more rational energy use. "There isn't anyone like Amory in Asia," says Christine Loh, Hong Kong's undersecretary for the environment, who has known Lovins for many years. That lack of "an energy geek with a very practical mind-set," as Loh characterizes Lovins, has meant that the adoption of more economically and environmentally rational energy policies has been slower than it would have otherwise been.

The technology is available for more energy-efficient buildings, using myriad new and higher-technology materials, better design, and more efficient operations. The advent of green buildings does not mean a new age of austerity. It does not mean hot, stuffy office buildings where workers swelter in Asia's never-ending tropical summers or shiver in its bitter northern winters. Green buildings can in fact be more pleasant than conventional buildings—they use more natural light, and they are kept at a more comfortable temperature. The use of interior green walls, made up of plants, produces fresher air and a more pleasant environment.

If the benefits are so apparent, why have builders and building owners been so slow to embrace efficiency? The answer to the riddle of why these opportunities are squandered lies in a mix of a conservative industry, inappropriate regulations, and misaligned incentives.[7] The building industry in Asia seems to be filled not with people like Amory Lovins but with more cautious people who follow rules carefully rather than trying to innovate. That is on the whole a good thing, for buildings are complicated and poor design can mean a catastrophic failure. But conservatism can stifle new kinds of design.

This reluctance to innovate is often coupled with the idea that change—such as increased energy efficiency—means additional expense. Cost is often cited as a reason not to build greener buildings, and it's true that the nominal price tag for a so-called green building could typically run 3 to 10 percent higher than that of a conventional building in the earlier days of the transition away from building-as-usual. But experience shows that actual additional costs of greener buildings typically fall quickly as everyone involved in the industry, from architects and engineers to contractors and building owners, moves to using better materials, techniques, and operating procedures as a matter of course.[8]

In Singapore, one of the countries where energy-efficient building is most advanced, good environmental designs are simply incorporated into all projects. Part of this reflects mandatory building codes and part of it generally accepted practices. Lim Chee Onn, who has been involved in all four of Singapore's large-scale government-to-government city-building projects in China since the mid-1990s, notes that planners think not in terms of green but simply in terms of good design, with environmental efficiency taken as an essential element. Leading-edge design will continue to cost more, as it typically does in any case. Yet even for those early adopters who paid prices up to 10 percent higher, it can take as little as six months (although sometimes up to seven years) to recoup this in the form of savings on the cost of electricity.[9] Higher electricity prices translate into faster payback periods. Countries like India, with its high electricity prices, are adopting green buildings faster than China, with its relatively low power prices.

Building codes are complex and, again with good reason, conservative. In Asia, these codes until recently have had little in the way of efficiency standards—also for a reason: many governments in Asia tend to be better at building and managing large, discrete projects, like power plants, than at enforcing their will at the local level through micropolicies such as building codes. Notably, without cooperation and commitment from frontline officials and those in the industry, China's energy-efficiency edicts will remain of limited effectiveness. China, for all its reputation as a strong state, has yet to convince local officials and those in the construction and building management industry that the issue of energy efficiency is linked to national survival. Indeed, green building efforts work best when there is a broad social consensus backing them, for the myriad decisions made while constructing and operating each building are subject to corner cutting and energy-efficiency equipment such as air-conditioning is of only marginal value if it is not operated correctly. Governments can play a major part in instituting change, but their efforts must be coupled with a popular consensus that greener buildings are necessary.

However, there are many hurdles to implementing effective government regulation—and also to inciting public interest in change. To change behavior takes incentives, whether in the form of higher prices for energy or greater regulation, but it is difficult to target incentives to encourage energy-efficient building in a frenetically short-term-oriented society such as China's, where developers try to get their capital returned as quickly as possible and most government officials cannot see beyond a single year's growth targets. And although the overall cost of electricity to a society typically is significant, the part that any individual person or company must pay at any one time tends to be fairly insignificant, at least measured as a percentage of an individual's income or a company's operating expenses. The money-saving impact of energy-efficient design or operation on electricity bills in any given month is trivial. Tenants in office buildings or commercial shops have little bargaining power to change what is in any case a fairly small part of their monthly cost, so they do not push for improved energy efficiency. Landlords who pass on the costs of higher electricity bills to tenants, either directly or indirectly, also have no incentive to change.

Even if popular pressure did mount, developers who build for sale would still have no incentive to change their energy-inefficient ways—at least not yet—because they are not required to disclose the cost of their poor efficiency to buyers. The cost of the energy used to make the steel, cement, or glass in the building is similarly invisible to buyers. However, developers and buyers who ignore energy-efficiency opportunities could be putting their assets at risk. In Europe, buildings that are not energy-efficient are in danger of realizing returns from both rents and sales that are below those of more energy-efficient buildings. In the United States, more energy-efficient buildings also tend to have lower operating costs, higher occupancy rates, and higher rent receipts and selling prices.[10]

~

As recently as 2010, green buildings in Asia were an exotic breed. Now, many major real estate developers have proclaimed their conversion to the green doctrine. Green building organizations have sprung up in most Asian countries, trailing a gaggle of certifications in their wake. The U.S. Green Building Council's platinum and gold Leadership in Energy and Environmental Design (LEED) ratings, or their local equivalents, are increasingly needed by building owners who hope to attract multinational tenants. LEED started in 2000 and is regarded as a global benchmark, although it has often been adapted by national green building groups to suit local climates and building practices, such as in India where water shortages have driven green building standards to focus on harvesting rainfall, reusing water, and consuming less water overall.

Some green communities are making LEED certification a foundational requirement. The 1,500-acre international business district in Songdo, the city discussed in chapter 3 that is being built from scratch near Seoul's international airport, has dictated that 80 percent of the development in the site must hew to LEED standards, the largest-scale LEED requirement in the world. As of mid-2012, Songdo had 13.7 million square feet of LEED-certified building space, and Asia as a whole had 441 LEED-certified projects representing 211 million square feet of floor space.[11]

The notion of LEED-certified buildings is now a fashionable real estate statement, but it is not automatically a guarantee of a more energy-efficient building. LEED is a point-based system that does not evaluate the actual operation of the building. It gives points for everything from site selection to bicycle racks, and it mandates the use of a variety of energy-saving materials and equipment. However, it is based on initial design, rather than on performance. Moreover, LEED takes as a benchmark roughly similar buildings and asks developers and owners to do just a little bit better in order to achieve LEED certification. The weaknesses of LEED were displayed when a 6,721-square-foot house in a luxury gated community on the outskirts of Las Vegas garnered the highest rating, LEED platinum. This prompted Kaid Benfield, who helped develop the LEED for Neighborhood Development organization, which looks at the broader urban context, to write a blog piece titled "As Good and Important as It Is, LEED Can Be So Embarrassing."[12]

So although compliance with LEED or other building energy-efficiency standards is a start, a real, thoroughgoing commitment is a different matter altogether. Too many companies still appear to have only a superficial commitment to the idea of energy efficiency, focusing on small changes but ignoring big-picture gains. And even though small changes can indeed compound to make a meaningful difference, the most successful green companies are those that embed the notion of efficiency in the corporate culture and that strive for a deeper process of commitment, rather than superficial compliance with an external code. The best companies tend to start slowly, experimenting and learning. There is usually a series of fairly easy steps that result in savings in energy, water, or other areas and that fairly quickly repay the initial investment. These savings and the general good feelings that result often start a virtuous circle with a more sustained focus on energy and resource efficiency.

One of the companies that appears to have internalized a thoroughgoing commitment to energy efficiency—both in its buildings and throughout its company—is Esquel. This privately held Hong Kong company that makes more than 100 million men's dress and polo shirts every year is a leader in a variety of sustainability initiatives. As a major manufacturer in a low-margin business, it has internalized the lesson that any cost savings will make it more competitive and potentially more profitable.

In the early 2000s, Esquel engaged Civic Exchange, a Hong Kong think tank, to help it understand energy efficiency and environmental sustainability. One of the areas that Esquel focused on was lighting, where many small changes can add up to big savings. First, company employees measured the brightness and removed unnecessary light tubes. Then they replaced traditional fluorescent light tubes with more energy-efficient ones containing electronic ballasts, which allowed them a power savings of 18 percent. The addition of high-efficiency reflectors in the light fixtures meant that lower wattage lights could be used. In all, the company invested $490,000 and recouped that amount in sixteen months. Other energy savings came from automatic door closers, temperature controls, improved ventilation, and more natural lighting. Air conditioners were converted to run on power from the company's captive power plant.[13] It has also taken impressive steps to improve efficiency in its use of water, which will be discussed at greater length in chapter 6.

Companies like Esquel are helping to bring energy efficiency into mainstream Asian business, through new and often striking initiatives and innovations. Japan and Singapore are leading the way. Japan has been focused on energy efficiency since the oil shocks of the 1970s, and buildings have become a more important part of that effort in the past decade. Singapore explicitly adopted a green energy plan for buildings only in 2005, but it has come further than any Asian country. Other economies, including South Korea, Taiwan, and Hong Kong, have adopted energy-efficiency standards in the past decade, although their effectiveness to date varies widely.

Singapore's regional leadership merits a close look. That country is making a concerted government effort to ensure that its buildings become progressively more energy-efficient by 2030. This reflects the ethos, discussed in the last chapter, that the small country must use its limited resources wisely.

~

Singapore started its green building program in 2005 with the launch of the government's Green Mark Scheme, a home-grown benchmarking standard that adapts international energy-efficiency practices to Singapore's urban, tropical environment. The aim of the Green Mark Scheme was to cut water and electricity bills, reduce the environmental impact of the built environment, improve the indoor environmental quality to make buildings healthier and more productive, and provide a road map for continual improvement. The Second Green Building Master Plan (launched in 2009) widened the scope, focusing not on the construction of new buildings but on retrofitting existing buildings.[14]

As it has done in a number of environmental areas, Singapore started out fairly modestly but with a long-range goal in mind; its two green building plans have been incremental steps toward a clear, common aim. Its initiatives are scientifically based and government led, but they typically have strong support from companies (especially those that stand to gain from

The exterior of the Zero Energy Building, a retrofitted building at the Building Construction Authority Academy in Singapore that serves as a test-bed for green technology. It is estimated to save S$84,000 ($66,000) a year in energy costs compared to a typical office building in Singapore. Photo credit: David McIntyre

new programs) and civil society (encouraged by media campaigns and community outreach). In this case, the overall goal is to have 80 percent of the island's buildings certified as green by 2030, with a target of S$1.6 billion ($1.26 billion) in annual energy cost savings.[15] To speed this process, the government has committed S$100 million ($79 million) to pay for up to 50 percent of the cost of retrofitting under its Green Mark Incentive Scheme for Existing Buildings. Already, the number of green buildings in Singapore has grown from 17 in 2005 to almost 1,700 in September 2013, totaling about fifty million square meters or 21 percent of the total building stock.[16]

In concert with these efforts, progressively more stringent laws and regulations are gradually enforcing energy efficiency. New regulations under the Building Control Act, which went into force in mid-2013, require building owners to submit building energy consumption data annually. As of January 2014, any new or retrofitted cooling systems had to adhere to minimum Green Mark standards, and all systems must be checked every three years for continued compliance.

Singapore has won international recognition for its efforts in the building sector. In October 2013, the Singapore Building and Construction Authority became the first government organization outside of Europe and North America to receive the International Star Award from the Alliance to Save Energy, a Washington, D.C.–based organization founded by Senators Charles Percy and Hubert Humphrey in 1977. Sharing the awards were U.S. Secretary of the Navy Ray Mabus, the Los Angeles Transportation Authority, and Whirlpool.[17]

Although Singapore provides a remarkable example of a government that's taken on the task of promoting green building countrywide, most other initiatives throughout Asia are spearheaded not by governments but by individual companies and organizations. The rest of this chapter is devoted to snapshots of some of the most innovative work being done by such entities.

~

Ronnie Chan, chairman of Hang Lung Properties, is a green building convert. Hang Lung is one of Hong Kong's most prominent developers, and it has won numerous design awards. In recent years, most of its investment has been in mainland China. It has major projects in eight Chinese cities, and it is focused for the most part on large-scale, integrated centers. Imagine Rockefeller Center, with its combination of office and retail space—then add apartment buildings, and multiply the size. These are, as Chan says, "huge, humungous" developments. Chan's strategy is to "build world-class complexes for long term holding in economically vibrant cities" in China.

Chan, who is also cochairman of the New York–based Asia Society, was initially skeptical of the idea of green buildings. But he saw that where Hang Lung's Chinese investments are concerned, green building will ensure continued economic growth. Energy-efficient buildings will not only reduce the need for China to build as many new power plants, but also Chan expects that the Chinese government will impose increasingly strict energy-efficiency standards on buildings. He wants to be ahead of policy, rather than trying to play catch-up. The approach also makes good economic sense for a developer like Hang Lung, which keeps properties in its portfolios, rather than selling them. More energy efficiency means lower operating costs for buildings. The savings can be passed on to tenants or kept by Hang Lung. The company's goal is to reduce overall electricity consumption 5 percent from 2010 to 2015.[18]

All of Hang Lung's buildings in China are LEED gold certified. (The gold standard is above silver but below platinum in the LEED rankings.) Hang Lung's commitment means, according to the company's 2012 Sustainability Report, that its buildings include a more energy-efficient envelope (the building's concrete-and-glass shell), systems to recover and use waste heat and water, and energy-efficient lighting and air-conditioning. The hope is that this will translate into significant energy savings. As an example, its Olympia 66 building in the northeastern Chinese city of Dalian expects energy savings of 14 percent and water savings of 45 percent over ordinary comparable buildings when it opens in 2015.

Hang Lung's approach exemplifies a pragmatic rather than a radically innovative approach to sustainability. Chan builds his buildings to the standard that the market will pay for—as evidenced by choosing LEED gold instead of platinum. "The first thing of course is that it is the right thing to do economically," says Chan, whose company took its first energy-efficiency steps in 2004 but didn't really embrace the concept until the end of that decade. "You have people who do one platinum building. That is not as half as significant as doing ten gold buildings." Chan says that the payback period on energy-saving initiatives—the time it takes to generate enough savings to pay for the investment—can be as short as about two years.

Although retrofits are more expensive and generate less of a return on investment, Hang Lung is also upgrading existing buildings to generate energy savings. It had retrofitted the air conditioner chiller systems in twelve of its Hong Kong properties by 2012 and will spend HK$148 million ($19 million) over the following decade installing the new chillers at ten more Hong Kong properties.[19]

Hang Lung is taking a long-term mind-set in its approach to green building. Governments want more energy efficiency, as do tenants, especially multinational corporations. "You have the government pushing from the top and your tenants or potential tenants pushing from below," notes Chan. "If you don't do it you may be sunk one day. The government may require you to retrofit. We are right now retrofitting some of our older buildings. . . . We know how expensive retrofitting is. We know we want to do it right from day one. The government is going to breathe down your throat or your tenant will."

~

The Swire Group is one of the few remaining colonial-era *hongs* still controlled by its founding British family. Established by John Swire as a Liverpool trading company in 1816, the group set up shop in Shanghai in 1866 and opened an office in Hong Kong four years later. This makes it one of the oldest companies in a region where few large businesses predate 1945.

It's not just its Cathay Pacific airline that would make the Swire of today unrecognizable to its founder. Merlin Swire, the latest of a long line of family members involved in running the business, now oversees an empire that includes one of the world's largest Coca-Cola bottling enterprises as well as a significant property development arm and a mixture of other businesses ranging from a venture with Campbell Soup to specialist offshore vessels servicing the energy industry.

Swire is an enormous group. Its various companies employ more than 120,000 people, and in 2013, it reported profits of more than $1.7 billion on sales of $6.6 billion.[20]

Swire's continuing prosperity, two centuries after its founding, reflects its ability to think for the very long term. So it is little surprise that sustainability has become a core part of the group's ethos in the past decade—although in its internal communications the company prefers to position the topic as part of a larger discussion of financial returns and risk management. "You talk about sustainability and people's eyes glaze over," says Philippe Lacamp, head of sustainable development at John Swire & Sons (Hong Kong). "You talk about energy efficiency and risk management and people say 'I've got it.' . . . Sustainability is about risk management on a larger scale." So to its managers Swire talks about risk and good business practices. In public, the company proclaims its commitment to sustainability at the beginning of its annual report. "Sustainability is key to our long-term approach. We recognize that sustainable development does not mean less profit. Rather it is an opportunity to increase efficiency. Our ultimate goal is for our operating companies to achieve zero net impact on the environment."[21]

Indeed, so thoroughly is the sustainability thread woven into the corporate fabric that Swire Pacific does not report on these issues separately but incorporates them into its central financial reporting. "If you want change in any organization you have to have accountability," says Lacamp, who has been with Swire for over twenty years. "Sustainability is already almost a hackneyed term. How do you make it a part of the management process? The example I use is health and safety. Twenty years ago health and safety was where sustainability is today. 'Oh, what a pain—the head office is after us. It is not a key part of the business.' Now it is unthinkable that you wouldn't have health and safety reporting statistics, spotting the trend lines. I saw that change when the operating companies had to sign off—the general managers and the CEOs. The moment that happened there was attention paid to the data. Through the reporting, we are getting to that point on sustainability."

Externally, Swire reports to the Carbon Disclosure Project and is a member of the FTSE4Good and Hang Seng Sustainability Indices.[22] More important are its internal reports. The company's sustainability committees—which document not only the environment and energy but also human resources, health and safety, and the supply chain—report to the group's risk management committee.

Swire concedes that it is many years from achieving its goal, first articulated in 2010, of contributing net zero impact on the environment.[23] In 2012, the company came up with a preliminary road map for achieving this goal, and already it's reporting on a range of measures from water and energy consumption to greenhouse gas emissions on a clearly transparent basis.

Swire Properties, which produces the second-largest amount of emissions in the group after aviation, is the only one of the group's five divisions that has come up with a hard numerical target to start moving toward net zero. It aims to cut energy use at its Hong Kong investment properties by 50 million kilowatt hours by 2016, using 2008 levels as a base. The company estimates that this will save about HK$66 million ($8.5 million) in annual electricity bills. This is the equivalent to turning off almost 60,000 standard 100 W lightbulbs running night and day for a full year.[24]

The impact of its green initiative is already apparent, as Swire Properties saw its property portfolio grow 7 percent in 2012 and yet saw its energy use drop 11 percent and its water use grow just 1 percent. It owns 28 percent more floor space than it did in 2001, but its energy use has gone up only 5 percent.[25] As explained in its annual report: "Tenants increasingly scrutinize the sustainability credentials of landlords and buildings, and Swire Properties aims to be at the forefront of sustainable development by designing energy efficient buildings through the innovative use of design, materials and new technology."[26]

In 2013, the Swire Group began using quarterly reporting to help it achieve its sustainability targets. Having set up a carbon desk in 2012, it has experimented with internal carbon trading—recognizing that its aim of being net carbon neutral will require significant offsets to balance emissions from Cathay Pacific and its property business. As a result of the discipline of comprehensively understanding its many businesses' environmental footprints, it is now uncovering possible new ways to generate carbon savings. Lacamp, for instance, sees its various agribusiness operations as a way to generate carbon credits. For example, flat roofs on its cold-storage facilities can be used for solar. "We are engaged with our operating companies. If they can come up with voluntary credits we can buy them ourselves. [Perhaps someday] the public can buy from private companies, so we potentially could sell those into a broader market—Swire-branded carbon credits."

Sustainability initiatives are a business tool for Swire. They help the company keep the social license to operate, a phrase that recognizes the danger to corporations if they lose public trust and support. "The social license to operate is a key driver for sustainability," says Lacamp. Its Coca-Cola bottling unit has what the company says is one of the world's most energy- and water-efficient bottling plants. Examples like this help it convince local governments and people that it is a responsible corporate citizen. Sustainability is not an element that is isolated from broader business activities. "We use that as a business driver," says Lacamp. "That puts us in a much stronger position."

This social license to operate is especially important for Swire as a hybrid British–Hong Kong–Asian company without any natural support from Hong Kong's postcolonial government. "We don't have a government that will stand up for us. Why should they? We are a merchant. There is no right for us to be here."

Sustainability principles serve an important internal purpose in helping Swire better manage risk and gain a competitive advantage. That is certainly the case when it comes to buildings. Like Ronnie Chan at Hang Lung, Lacamp believes that there will be increasingly stringent energy-efficiency standards. Given that Swire's buildings are meant to last fifty or more years, it makes sense to stay ahead of expected legislation requiring mandatory retrofits to increase energy efficiency. "We think there will be laws requiring retrofits," says Lacamp, and building to high standards is more cost-effective than retrofitting. "I want that bar as high as possible because I want to knock those guys who are not behaving out of business.... The sustainability component and the commercial imperative overlap. That is my sweet spot."

"When you work for a company that has been around as long as Swire, it doesn't take much for us to say 'you need to think longer term,'" says

Lacamp. "We can get people to step back from today's spread sheets and say 'What do you need to be thinking about to be around in 20 years?'"

~

In India, outsourcing giant Infosys has embraced sustainability out of necessity. It operates in a country where both water and electricity are expensive and often unreliable. Infosys was one of the pioneers of the Indian outsourcing boom, and now is listed on the New York Stock Exchange. The company received a big boost with the Y2K work at the turn of the century and capitalized on the growing popularity of outsourcing in the decade that followed. With most of its more than 150,000 employees based in India, where labor is relatively inexpensive but electricity and water are not, the company recognizes that in order to keep growing it must have a comprehensive plan for energy and water efficiency.

Infosys has pledged to reduce per capita energy consumption by 50 percent and to power its Indian operations with renewable energy by 2017. Infosys now gets about 22 percent of its power from renewable sources and hopes to have the number up to 50 percent in the near future, says the company's cochairman Kris Gopalakrishnan, one of seven people who founded the company in 1981. "We do not have the luxury of time anymore," adds Rohan Parikh, head of the company's environmental initiatives. "Businesses, government and society have to re-prioritize, refocus and challenge the fundamental framework of our current economic model. We need leaders who have the courage and conviction to take unreasonable goals. We have taken the goals to become carbon neutral, reduce our per capita energy consumption by 50 percent and power our entire Indian operations by renewable energy."[27]

Since 2008, Infosys has cut its per-person use of water 40 percent and electricity almost 50 percent, at no additional cost. The company's Energy Performance Index (EPI), which measures kilowatt hours consumed per square meter of space, is 85, impressive when compared to the EPIs of similar buildings in India, which average between 200 and 400.[28]

As with others, Infosys has found that the more it measured, the more it saved. It started by metering—and thus measuring—electricity use in each of its buildings, followed by submetering of major systems, such as the lighting, computer, and air-conditioning systems. The economic case for replacing equipment that is more than five years old is often compelling. Infosys says that the return on investments is three years for air-conditioning plants and just over four years for uninterruptible power systems. Gopalakrishnan points to design improvements that have helped the company save $7 million in electricity costs. "That's real money that has been saved," he says. Many of these are small tweaks. For example, to reach the goal of using 100 percent natural light for work spaces, Infosys moved conference rooms and managers' offices to buildings' cores, leaving natural light for rank-and-file workers, with the lesser-used rooms windowless. The company also put the window ledge at the top of the roof about one foot lower, angled it, and attached reflectors so that more natural light floods work spaces. Currently, some 60 to 80 percent of employees work with natural light.[29]

~

Jamshyd Godrej had high hopes when he conceived of building India's first green building as a way of showcasing the promise of environmental technologies in India. But even Godrej was pleasantly surprised that the U.S. Green Building Council named the CII Sohrabji Godrej Business Center the world's most energy-efficient building of its kind when it was built in 2003, awarding the center one of the first platinum-level LEED ratings. The demonstration building, in the southern city of Hyderabad, was built by an unusual partnership with the support of the U.S. Agency for International Development (USAID), the Confederation of Indian Industry (CII), and the Indian government as well as a Godrej family foundation.[30]

Godrej & Boyce is not only one of India's largest and oldest business groups but also one that occupies a unique place in India's struggle for independence. The company was founded in 1897 as a lock manufacturer by Jamshyd Godrej's grandfather at the urging of Mahatma Gandhi, as part of the burgeoning economic self-sufficiency movement that became a key part of the country's independence struggle. The Godrej family is Parsi—ethnically Persian Zoroastrians who came to India over a millennium ago. The Parsis are a small but influential group that long has played an important role in Mumbai's commercial and industrial life. Even today, Godrej & Boyce remains a company with a strong commitment to broader issues of importance to Indian society.

Godrej's headquarters and many of its manufacturing facilities are located on a vast campus not far from the Mumbai airport, well north of the original city center. The campus is 3,000 acres, almost four times the size of New York's Central Park, although 2,300 acres are set aside as a protected mangrove reserve.[31] In a metropolitan area of close to twenty million people, the size alone would make Godrej & Boyce's facilities unusual; that they have the largest mangrove swamp in the area may make them unique in the world. A new 1.4-million-square-meter building incorporates innovative features such as a roof landscaped with low-water grasses that harvests rainwater and keeps the building cooler and an organic waste converter that produces manure to be used in landscaping. All of this testifies to the unusual long-term vision of the Godrej family, a vision that is reflected in its approach to green buildings.

When I visited the campus at the end of 2013, Anup Mathew, the head of Godrej Construction, took me first to a showroom that displays the products that its more than ten thousand employees make. There are coffeemakers and forklifts, locks and bank vaults and safes, and furniture and rocket propulsion components. In 1958, Godrej & Boyce was the first Indian company to make refrigerators, and today its appliances also include air conditioners, microwaves, and washing machines. Across the street, in the mangrove swamp, three massive containment vessels are being readied for shipment to the Reliance Industries petrochemical facilities in Rajasthan, their cargo including some of the industrial products from a group whose annual sales in 2013 were more than $3.3 billion.

Jamshyd Godrej has had an impact not only on the sustainability of the company he runs but also on the environmental goals of his nation, where he is regarded as the father of green buildings. The Indian Green Building Council, which Godrej helped found, has registered 2,362 green building projects with a total of 1.8 billion square feet of space.[32] However, of these, only 447 buildings, or about 1 in 5 of those registered, have been certified. Certification is one step—but what really matters is the actual performance of the buildings. At Godrej's urging, the council is trying to get all certified buildings to report their actual performance. Otherwise, the rating system is in danger of becoming a meaningless seal of approval that doesn't accomplish the goal of saving energy. "There are a lot of architects who tell clients that their buildings are green," says Godrej. "Architects have a way of getting away with these things. If you are not building it to a standard, how do you know? I want every single rated building to be totally transparent. They must be completely transparent with their numbers on our [Green Building Council of India] website, on a quarterly or annual basis." Godrej recognizes that we need to move beyond simple measurement and look qualitatively at how well buildings serve the people who use them.[33]

This ethos of moving beyond merely checking off items on a list in order to meet green building standards has permeated the company. At Godrej & Boyce, Rumi P. Engineer—a third-generation, yes, engineer—took me on two tours of the company's facilities in 2013 and spoke about the need to "go beyond certification to commitment and conviction." From siting its new apartment buildings on an east-west axis so as to avoid the worst of the tropical sun to installing shaded overhangs and high-efficiency glass, it's clear that Godrej & Boyce architects and designers are trying not only to maximize energy efficiency but also to enhance the design aesthetics and the enjoyment of the apartments. Costs are about 5 percent higher for these green apartments, but the extra expense is worthwhile for the company, as they sell for 25 to 30 percent more than comparable ones.

In India, sustainability has been forced on companies and the country by necessity. Water is in short supply, and electricity is unreliable and expensive. Mumbai has put restrictions on the use of water in the last four or five years, so Godrej & Boyce is trying to make its campus water-neutral. Toilet flushing and landscaping at the well-manicured campus use gray water. A company-owned sewage plant treats this water, and it is used in making concrete.

Godrej & Boyce has adopted a number of good practices—it individually meters as much of the building as possible so that it can track major consumption points. Each elevator and each chiller (for the air conditioner), for example, is metered. This sort of submetering allows more careful monitoring, which in turn allows the company to keep energy consumption for the building at 135 kW per year per cubic meter, impressive when measured by both international and domestic standards.

"We are good but we can do better," says Engineer. The sort of continuous desire for improvement that I saw at Godrej & Boyce typifies sustainability leaders. There is no magic solution, no easy answer, to sustainability issues. Buildings and cities are complex, evolving systems. Godrej & Boyce underscores the need to have leadership from the very top, which Jamshyd Godrej has provided. Equally important is a systematic process for ensuring that this top-down vision is translated into action. The guidelines for green buildings do that for Godrej & Boyce. It is equally impressive to see how front-line staff try to go beyond those guidelines, solving the environmental issues just as they would other business problems.

Leading companies often set themselves new challenges. Swire tries to report under the demanding Global Reporting Initiative (GRI) standards, and in 2013, Godrej & Boyce was for the first time engaged in a project to try to calculate its greenhouse gas emissions and its carbon footprint. Sustainability is just part of good ongoing business operations.

~

When construction on the Shanghai Tower topped out in August 2013, the building was second in height only to Dubai's Burj Khalifa. This landmark of China's arrival may herald not just the advent of even taller buildings but also an era of more environmentally benign structures. The Shanghai Tower was designed to reduce wind loads by one quarter, allowing it to reduce structural steel by one quarter as well, saving some $58 million. Wind turbines provide supplementary power for the skyscraper. Geothermal energy helps power the tower's heating and air-conditioning. A striking double-layered glass facade, with the inner core of the building proper completely separated from its outer skin, allows enough space for nine separate public atriums with restaurants and viewing areas. The dual-skin design functions as a sophisticated form of insulation, further cutting the energy load. The building's designer claims that energy consumption will be reduced 21 percent, compared with a typical similar-sized building, and that more than three dozen energy-saving technologies will cut the building's carbon footprint by 34,000 metric tons a year.[34]

Impressive as the Shanghai Tower's features are, trophy projects like this will not solve the energy problems posed by Asia's cities. That will take commitment of a sort that few countries have shown.

The slow pace at which green buildings are constructed in Asia is due to a lack of leadership and a general sense that business-as-usual is good enough. For those who believe that green buildings' up-front costs are significant but that benefits are diffuse and long term, making buildings more energy-efficient just doesn't seem to be worth the trouble. With a few exceptions, governments have not required more energy-efficient buildings; real estate developers have not seen the economic rationale; and end users, either purchasers or tenants, have not cared enough to demand better. Companies with the most incentive to change are those that build, own, and occupy their own buildings. Godrej & Boyce, Infosys, and Esquel are good examples in this category. Companies that both develop and operate buildings, such as Swire Pacific and Hang Lung, invest in energy efficiency because it holds costs down and enhances their brand.

For decades, Asian developers have counted on making money by building cheap buildings with short lives. That strategy, which depends on rapid increases in property prices, is likely coming to an end. Economic growth rates in China and much of the rest of the region are coming down. Whether looking at property prices as a percentage of income or comparing them to global norms, there's likely to be more modest property price appreciation in the decade ahead in the region's leading cities. That means well-built, well-maintained, energy-efficient buildings will fare better. There is a challenge here but also an enormous opportunity: to build more energy-efficient buildings, to build them to last longer, and to use less energy in construction and reuse more of the materials. Building energy efficiency is the energy equivalent of a free lunch, a benefit that costs nothing—in fact, it pays those who invest in it.

This opportunity must be seized—and quickly. There are many individual companies doing outstanding work. The notion of green building is becoming mainstream. But government must push harder and faster to work with business to upgrade building efficiency standards.

5

Asia on the Move: *Cars and Trains*

> Electrified transportation not only solves China's problem but solves the world's problem.
>
> —WANG CHUANFU

From rickshaw to Rolls-Royce: Through much of the twentieth century, Asian streetscapes were filled with carts pulled or pedaled by people, but today Asia's car-clotted cities exemplify the enormous material progress of the past few decades and the environmental challenges ahead. Nowhere is this more true than in China. When China's economic reform began in 1978, bicycles vastly outnumbered cars on the streets of Beijing, with the few cars usually carrying government officials. Now, the capital is circled by six ring roads; yet even with this huge expansion of the traffic network, the cars often move more slowly than the remaining bicyclists.

China today is the world's largest automobile market, and Chinese buyers account for one of every four cars sold in the world. In 2012, a staggering 17.9 million private cars were sold in China, more than twice the U.S. figure of 7.6 million. Auto sales around the region tell a similar tale. In 1980, less than one car in ten in the world was in Asia. Today, it is one in five, and by 2030, the figure is expected to be almost one in two.[1]

Along with the skyscraper, the automobile is one of the most visible icons of affluence and progress. Yet as Asia becomes richer, the automobile has also become a symbol of so much that is wrong with the twentieth-century narrative of development. Private cars hold out the promise of freedom, yet their popularity quickly imposes the tyranny of the traffic jam.

Asian cities are choking on their own success, with chronic traffic snarls both an inconvenience and an economic drain. The pollution that comes out of the tailpipe of traditional combustion engines is a significant contributor to a regional health emergency; air pollution is responsible for about two million premature deaths a year in East and South Asia. Cars are one of the largest contributors to CO_2 emissions. Today, transport accounts for 23 percent of global CO_2 emissions, and road transport accounts for about three quarters of that. Asia's cities need more comfortable, convenient, and environmentally sustainable transport.[2]

It is possible, of course, that Asian cities could simply become more polluted and more congested. The optimistic view, however, is that a wealthier, more aware, and more vocal citizenry, coupled with governments that are attuned to their demands, will gradually nudge Asia's cities toward more sustainable alternatives, whether in the form of electric vehicles or investment in better and more extensive mass transport systems.

However, the health crisis caused by toxic pollution, the enervating traffic congestion, and the carbon emissions alone will not provide the catalyst for this change. It is worries of a more threatening sort—the fear that energy supplies could be cut off or that prices could spike—that are pushing governments to come up with policies to reduce reliance on the internal combustion engine. Like the United States, China is finding that its hunger for oil and other natural resources drives its foreign policy, especially in Africa and the Middle East. Uncomfortably, from a Chinese perspective, it remains dependent on the U.S. Navy to guarantee its supplies of oil. In 2013, China spent about $250 billion on imported oil, much of it used for its expanding auto fleet, not far short of the $439 billion in total exports it sent to the United States.[3] "Energy security is very crucial right now for Chinese," says Wang Chuanfu, the chairman of electric-car manufacturer BYD, echoing a common concern of Chinese executives and officials. "If ports are blocked there would be no oil in China. Government control would be unstable."

This desire for energy security is perhaps the single most important factor behind China's concerted embrace of electric vehicles. "Most people don't understand why China needs to develop the electric vehicle," says Wang. The issue of energy security "is ten times more crucial than the environmental issue," he says. If China can ensure that electric vehicles are sold widely, "the environment shouldn't be an issue anymore." China shares its energy security predicament with its neighbors. Most East and South Asian countries are significant net oil importers, and imports have been growing in step with the growth in car ownership.[4]

Asia's most populous countries—China, India, and Indonesia—are among many in the region that compound their dilemma by subsidizing oil consumption. Fuel subsidies are politically popular and help keep down the reported inflation rate, as fuel tends to be overweighted in inflation indices. But subsidies are costly and contribute to higher consumption, making it more difficult for cleaner fuel alternatives to compete. Globally, the International Monetary Fund (IMF) estimates that fuel subsidies cost an extraordinary $1.9 trillion, equivalent to 2.5 percent of global economic output or 8 percent of total government revenues.[5]

Yet governments are understandably reluctant to reduce or remove subsidies. Indonesian President Suharto's decision to roll back subsidies (under pressure from the IMF) contributed to his 1998 overthrow. Fifteen years later, Indonesia's partial removal of subsidies in June 2013 provoked protests in Jakarta and other cities; even with the cut in subsidies, they still will cost the country about $20 billion a year, equivalent to some 2 percent of Indonesia's GDP. Following the price hike, Indonesia's fuel still costs only about $0.66 a liter ($2.50 a gallon), far less than global prices.[6] Greater engine efficiency and a move to electric engines offer a politically palatable way to move away from fuel subsidies.

For all these reasons—environmental concerns, energy security, costly subsidies, growing wealth, and the sheer number of people—Asia could be one of the first places in the world where the traditional internal combustion engine gives way to an electric-gasoline hybrid or an all-electric car on a large scale. This change gives Asian companies a chance to play catch-up with their larger global competitors at a time when the future of automobile technology—be it a hybrid gasoline-electric car or a pure electric car or perhaps a different technology altogether—is less certain for the industry than it has been at any time since the internal combustion engine rose to dominance a century ago. Whether they are making key components, especially electric-car batteries, or finished vehicles, new businesses are opening up as the old order is shaken. A new generation of green cars is as likely to come from Asia as from the United States.

Asian companies are pursuing a kaleidoscope of transportation innovations. China's BYD is using its battery technology, developed for mobile phones, to catapult into electric cars, taxis, and buses as well as large-scale storage devices for solar and wind power. Its home-grown competitors include Tianjin Lishen Batteries (a company that, like BYD, enjoys significant state backing). Batteries have been a major area for Asian companies; South Korea's LG Chem makes the batteries for the Chevy Volt, and China's

Wanxiang Group provoked bitter protests from American conservatives when it bought bankrupt U.S. battery company A123. Boston Power, a U.S. company that produces lithium-ion batteries, received a $125 million investment from China's GSR Venture. Other automakers are funding research into more speculative technology. South Korea's Hyundai is working on the less-proven hydrogen fuel-cell technology and in mid-2014 became the first major auto maker to see fuel-cell vehicles into the U.S. market.[7]

"Electric vehicles represent a revolution in the auto industry," says Tianjin Lishen's President Qin Xingcai.[8] "Of course, it won't take place overnight. It may take 10, 30, or 50 years, but the trend is very clear." Companies around the region agree. India's Mahindra bought the pioneering electric-car maker Reva, and one of the world's largest automakers, Toyota, has already demonstrated extraordinary success with its Prius and other hybrid vehicles. Toyota's hybrids combine an electric battery–powered motor with an internal combustion engine, resulting in dramatic gains in fuel efficiency. The Prius was introduced in 1997, and by September 2014, the Japanese automaker's cumulative hybrid sales had passed seven million, including about two and-a-half million in North America. Toyota's wager that the future is in hybrids, rather than all-electric vehicles, has paid off: about 40 percent of its Japanese sales and 15 percent of its total global production consist of hybrids. "Regulatory push cannot, on its own, induce consumer pull," says Satoshi Ogiso, one of the developers of the Prius hybrid, implicitly referring to the difficulty that all-electric vehicles have had in gaining consumer acceptance, despite government regulations and incentives.[9]

This is a time of technological ferment in the automobile industry as both new and established automakers try to develop more fuel-efficient technologies. Nowhere is this more apparent than in Asia. The challenges and the opportunities presented by electric and hybrid cars are obvious in India and China. With more than one third of the world's population living in these two countries alone, the gasoline-engine-powered economic model of the United States or Europe is unthinkable. Fortunately, both China and India have deep technical and engineering skills and a strong local automobile manufacturing base. BYD and Mahindra Reva—which are discussed in greater detail below—have substantial financial and managerial resources, although they remain bit players in the global automotive world. They are emblematic of the engineering-led innovation emerging from scores of companies in Asia as engineering skills are increasingly being used to create more environmentally benign products.

China has been particularly aggressive in trying to make the transition away from gasoline-powered engines. In late 2008, as the financial crisis ricocheted through the world, China adopted a massive stimulus program, earmarking $586 billion for a variety of projects and picking electric vehicles as one of seven pillar industries.[10] At the time, China set a goal of putting 500,000 electric vehicles on its roads by 2015. In 2012, it upped the ante, establishing a target of five million electric vehicles on the road by 2020. China's determination to jump-start the electric-vehicle industry demonstrates both the power and the capacity of the Chinese government, working in concert with companies and research laboratories.

Yet China's experience also shows the limits of state planning, in what is still something of a command economy, for these efforts so far have fallen short. Despite heavy investments and generous subsidies, only 14,604 electric cars and 3,038 plug-in electric hybrids were sold in 2013, according to the China Association of Automobile Manufacturers. Most electric-vehicle sales in China have been made to cities buying electric buses and taxis.[11] In China, as elsewhere, higher prices (50 to 100 percent more than conventional cars), coupled with consumer worry about running out of battery power and the lack of charging stations, have slowed sales. It is uncertain if we are too impatient for adoption of electric vehicles or if critics are right that this will prove to be a dead-end technology.[12]

Tianjin Lishen's CEO Qin Xingcai is a realist in this brutally competitive, fast-moving, capital-intensive, and technologically complex industry, saying, "When it comes to an industrial revolution, some players will win and some will be sacrificed." What is certain is that companies such as BYD and Mahindra Reva illustrate how entrepreneurial zeal and significant financial and managerial resources are injecting new energy into the auto industry.

~

Bangalore bills itself as India's answer to Silicon Valley. With multibillion-dollar outsourcing companies like Infosys and foreign multinationals like Hewlett-Packard nestled on leafy corporate campuses in the city, the comparison has a certain superficial plausibility. Yet no one who has visited both places would be in danger of confusing the two. The shorthand comparison is an attempt to make a very different country comprehensible to Americans.

Bangalore, India's third-largest city, is home to more than ten million people. This population—larger than New York City's—is jammed into a substantially smaller area, with its crowded feel accentuated by the absence of high-rise buildings, the lack of a subway system, and an overburdened road network. The greenery bequeathed to the city by the British allows Bangalore to call itself the Garden City of India, but the fast-growing population, which has grown twenty-five times since 1941 and, more recently, doubled between 1991 and 2011, is outstripping civic services. Where Silicon Valley has shopping malls and expressways in the shadow of the Santa Cruz Mountains, Bangalore has traffic that typically moves at just six miles an hour—jogging speed—and lurches from one jam to another.

The founder of one of the world's oldest and most innovative electric-car ventures knows both places well. Born in 1971, in a Bangalore that still retained its leafy charm, Chetan Maini grew up building remote-controlled cars and planes and, later, go-carts.[13] His father was in the electronics business and encouraged his son's interest in tinkering. From Bangalore, Maini went to the United States to study. In the past three decades, he has shown extraordinary vision and leadership in electric vehicles, but he has also run

Chetan Maini, founder and CEO of Reva, an Indian electric-car company that was acquired by Mahindra & Mahindra in May 2010. Photo credit: © AP/Aijaz Rahi

up against the hard reality of clean-tech start-ups—in 2010, he surrendered control of his company to one of India's largest conglomerates.

Maini's foray into electric cars started when he was a nineteen-year-old undergraduate engineering student at the University of Michigan in Ann Arbor in 1990. As part of a university team, he helped build the electric car that won the General Motors–sponsored Sunrayce USA, an 1,800-mile electric-car race from Disney World in Orlando, Florida, to the General Motors Technical Center in Warren, Michigan. The average speed was twenty-five miles per hour. Next, the team went to Australia, where it placed third in a grueling trek across the continent, behind a winning Swiss team and a far more lavishly funded Honda team. "It was an eye-opener," remembers Maini of the 1,900-mile cross-country race across the unforgiving Australian outback. "The fact that you could cross a continent on sun energy was amazing." So, too, was what could be accomplished by a scrappy team with little funding but an innovative strategy. "We had never built a car before, and we were racing against large teams. The strategy was different. Doing something new was very exciting. We were not the highest performance car, but we had better strategy."

A better strategy has remained Maini's guiding principle, but he has nonetheless struggled as an electric-vehicle pioneer. Buoyed by the excitement of the team's accomplishment, Chetan—as his staff calls him—decided that his future was in electric vehicles. He has been on the bleeding edge of the electric-car world ever since. He moved to California to set up Reva, earning a master's in engineering at Stanford University along the way. After California decided not to adopt a requirement that electric vehicles make up 2 percent of the state's fleet, Maini returned to India with his wife, a Michigan native, and began building the company there.

In 2001, Reva started producing cars in Bangalore. This gave Maini the distinction of selling some of the world's first road-legal electric cars—rather than, say, electric golf carts. It wasn't an easy journey. Maini recalls that when he approached investors, the attitude was "Crazy company in India—who is going to invest?" Everything, he remembers, "was a challenge," from winning government approvals and certification in an industry that was built around gasoline-powered engines to building a trusted base of manufacturing suppliers. The company had some modest success—more than one thousand of its cars have been sold in London, for example, where electric vehicles are exempt from the congestion tax. Still, by the time it launched its second-generation model—the e20—in 2013, its total sales numbered fewer than five thousand.

What the company and its founder have lacked in sales, they've made up for in accolades. *The Economist* awarded Maini its Energy and Environment award in 2011. In 2013, Reva ranked twenty-second on *Fast Company*'s list of the world's most innovative companies—a ranking it shared with Tesla, perhaps the world's best-known electric-car company.[14] Tesla sells cars priced between $59,000 and $107,000; Reva's were introduced at about $11,000; subsequent price cuts made the base price about $8200 in 2014. "Our two companies come at the same technology from completely different ends," says Maini. "There is a complete contrast, but there is something common about the efforts." Tesla has established a niche as a luxury-car maker. Reva is trying to find a place that is in the more price-sensitive middle of the market.

Electric vehicles have struggled everywhere to gain market acceptance. A combination of high initial costs and limited range—and fear of the car running out of power—has held back sales around the world. Reva was no exception, and the global financial crisis made a difficult situation worse. In May 2010, Maini sold a majority stake in Reva to the Mahindra Group, an Indian conglomerate whose businesses are as diverse as software and hotels and whose various vehicle-making subsidiaries produce everything from motorcycles to buses. The firm makes more tractors than John Deere—or any other company in the world. Mahindra, with yearly group sales of more than $16 billion and 155,000 employees, brought the sort of scale that Reva needed. It had a network of dealers, financial strength, and manufacturing know-how, and it also had other vehicles that could be adapted to use the Reva electric technology.

Mahindra's backing allowed Reva to build a new factory, one that company officials proudly note was the first factory to win a platinum rating (the highest level) from the Indian Green Building Council. The facility, spread out over 4.2 acres, has a capacity to produce thirty thousand cars a year (although current production is well short of that). Most of the factory's lighting is natural, and the little electric light that is used is mostly energy-efficient LED's. Maini rejected initial designs to build a factory with conventional air-conditioning; instead, an innovative design and a specially insulated roof keep the temperature on the factory floor comfortable. During my visit on an unseasonably hot, late April day, the thermometer read 99 degrees on the street, but the factory remained cool. The factory also practices rainwater harvesting and uses solar power for one third of its electrical needs.

In a country where both water and power are in short supply, this approach makes good business sense. "People realized that green didn't have to be more expensive. Green is about making things more efficient," says Maini over a lunch of dal, roti, and other home-cooked food on the roof of his seven-story office building overlooking Bangalore. "It becomes second nature—as you get cost reduction, good styling and design don't have to cost more.... It's about a shift in mind-set." Elon Musk's Tesla makes a luxury car that happens to be electric-powered, and Maini's Reva strives for frugality.

With the e20, Maini and his team rethought the whole car-making process so they could produce a vehicle inexpensively, with relatively little labor, in small quantities, and in a variety of locations. Lightweight, dent-proof plastic panels are sourced from an Austrian supplier that is 90 percent powered by renewable sources. These pieces are fastened together almost like a set of LEGO blocks. This modular and thrifty approach extends to the whole assembly process. A single car can be put together with fewer than ten hours of labor, using no robots and little automation. The company claims that the cost of setting up a manufacturing facility is one tenth that for a typical plant.[15]

The fully automatic four-seater is powered by a lithium-ion battery compact enough to leave room for a spare tire. Five hours of charging provide the e20 with its 100-mile range. The battery also gains a small charge every time the vehicle's brakes are applied. Fittingly, for a man whose team strategy in races was consistency rather than speed, the e20 is not fast. The car has a top speed of just 81 kilometers per hour—meaning that drivers are unlikely to get a speeding ticket in metropolitan India, where the speed limit is 80 kilometers, or 50 miles, per hour. Faster, longer-range versions are on the way, part of plans that call for a higher-performance model to be exported to Europe.

Despite the e20's frugal design, it has some higher-end features that are intended to appeal to the target market of environmentally minded urbanites looking for a second car. A partnership with communications giant Vodafone has produced a Bluetooth-enabled system that allows the car to be locked and unlocked remotely. The air-conditioning can be switched on remotely as well, allowing drivers to open the door to a cooler car—a benefit in India's extreme climate.

The Bluetooth feature also enables the driver to unlock an emergency power reserve if the car runs out of juice. This reserve supply will power the vehicle for up to six miles, hopefully far enough to get home or to the nearest charging station. Maini and his team claim that this so-called REVive feature is a world-first and will assuage one of the biggest fears of potential electric-car buyers—that they will run out of power. Reva, like other automakers, thought of adding more power to address this problem, but that would mean a bigger battery, which would in turn add to weight and cost. "We were answering the wrong question," says R. Chandramouli, the company's chief of operations. "It was not 'How do I know I am running out of power?' but 'When I run out of power, what is happening?' We answer the question psychologically first, not technically. It is anxiety reduction."

An electric car is a tough sell in a country that suffers chronic power shortages and has had the largest blackout in world history—more than 600 million people across the northern half of India lost power in 2012. As a result, Reva invented a solar garage costing $2,000 that allows e20 owners to charge their car without having to go on the grid.

Reva is a very tech-oriented company, with 35 percent of its workforce in research and development. It hopes to use its intellectual property (IP) prowess to create competitive barriers and supports these efforts internally with a dedicated IP committee that reports to the board of directors. So far, the results have been promising; the company holds fourteen patents, with another thirty-six inventions in the patent application process. Another aspect of its business plan is the licensing of key technologies to other companies, including its drivetrain, its battery technology, and its electronic information system—notably, its ability to perform remote diagnostics and data analytics as well as its ability to tap into a power reserve with REVive.

More broadly, Reva is encouraging consumers to rethink what they want from a car. It enlisted the eclectic Amsterdam-based advertising firm Strawberry Frog to do a series of provocative commercials with the theme "Ask" that challenged Indian consumers to think and ask about how a growing embrace of materialism will affect them, their country, and the planet. Coming from an advertising firm that caters to major clients—Jim Beam, LG, and Morgan Stanley, among others—these were high-end commercials designed both to promote the switch to a sustainable product and to underscore the overall quality of the e20.

Maini wants the e20 to be part of a broader transformation, not only in how products in India are made but also in how they are perceived. "People thought if it came from India, it was produced in a sweatshop," he says. The reality, Maini notes, is that "India is a country full of islands of excellence next to places of poverty and corruption." The advantage of India and Southeast Asia is the constant drive to innovate based on cost, he adds. "This is a company that is driven by ideas and technology. That is an important difference in how we see ourselves."

Still, government policy will be key to the industry's transformation, and Reva is counting on government support to strengthen the industry. In January 2013, India announced a national electric-vehicle program, which will be implemented by the National Council for Electric Mobility. Mahindra Chairman Anand Mahindra is a member of this body, and Maini is on the council's implementing board. Reva calculates that the estimated national investment in the five or six years following the inauguration of the board will be Rs 233 billion ($3.8 billion), with about Rs 140 billion ($2.3 billon) coming from the government. The government has talked of cumulative sales of six to seven million electric vehicles by 2020, with direct subsidies planned for both two-wheelers and four-wheelers. However, previous plans to implement these types of subsidies have been repeatedly delayed; it is not yet clear if the country will act on the program as currently envisioned.

Maini is a dogged visionary. Someone less committed to the idea of electric vehicles might well have given up long ago. After all, Maini has been at this for decades—he got involved with electric cars at the same time Tim Berners-Lee was developing the World Wide Web. The tie-up with Mahindra gives Reva access to capital and managerial knowledge on a scale that Maini could not achieve on his own. Yet it is still unclear whether or not Reva can achieve commercial success. Electric vehicles depend on subsidies to narrow the price gap with conventional gasoline-powered vehicles—subsidies that, even when promised, aren't certain—and manufacturers must fight consumer perceptions that the vehicles are not reliable. Maini has justly won accolades for his vision, and he deserves credit for his ability to sell his car in Europe as well as India. Maini is only in his mid-forties, so he still has time to achieve mass scale and see his vision vindicated.

~

During the darkest days of the global financial crisis in late September 2008, less than two weeks after Lehman's collapse, Warren Buffett's Berkshire Hathaway–controlled MidAmerican Energy Holdings spent $230 million for 9.9 percent of little-known Chinese batteries-to-autos conglomerate BYD. The Buffett investment focused attention on BYD and its energetic founder, Wang Chuanfu, who had parlayed his success as one of the world's largest battery makers into a head start in the race to build electric cars for the world's largest auto market. In 2010, *Bloomberg BusinessWeek* put BYD

American investor Warren Buffett (left) and BYD Chairman Wang Chuanfu attend a news conference in Shenzhen, China, on September 27, 2010. Buffett's backing gave the Chinese battery and automaking tycoon a huge boost. Photo credit: © AP/ Kin Cheung

at the top of its Tech 100 list, and *Fast Company* put BYD sixteenth on its list of the world's most innovative companies, sandwiched between Spotify and Cisco Systems. *Fast Company* said: "The Chinese car-and-battery maker already beat GM, Toyota, and Nissan to market with the first plug-in hybrid. Now the 15-year-old company is on the verge of doing the same with its all-electric full-size E6."[16]

After the Buffett purchase, the stock soared almost tenfold. But for investors, owning this stock has been more like riding in a roller coaster than a luxury sedan. In 2011 and 2012, it plunged, though even at its lows it was about 50 percent higher than Buffett's entry price. Improved performance in 2013 saw the stock double, putting its price at about three times what Buffett paid.

BYD is one of a breed of fast-growing Asian companies founded by and still very much a product of the vision of a charismatic founder. Wang

Chuanfu is a man with a dream, an engineer who bursts with ideas. With its fast-shifting strategy and experimentation with different sales approaches, his is the type of company that isn't common any longer in the West but that often proves successful in fast-growing economies with a kaleidoscope of shifting opportunities. Starting in 1995 as a battery maker, today BYD is one of the world's largest makers of nickel hydride and lithium-ion batteries, counting companies such as Amazon, Energizer, and Phillips as customers. It is also the world's second-largest maker of mobile phone cases, with clients that include Nokia, Samsung, Motorola, and Apple. It is a sprawling company, with businesses ranging from solar farms to mobile battery-powered energy units, from keypads to phone chargers to LED lights.

Although it was the Buffett move that brought Wang into the international spotlight, by the early 2000s he had already risen to some prominence as a battery maker. *BusinessWeek* named him one of the "Stars of Asia" in its 2003 list of Asian leaders. In the same year, BYD bought an unremarkable carmaker in Xi'an, a city far from China's booming eastern coast and that was China's capital during the ancient Tang dynasty. Xi'an is better known for its famed terra-cotta warriors than for its manufacturing prowess. But Wang's interest was in getting auto production technology—and a license—in order to start building electric vehicles. This 2003 acquisition began BYD's move into the electric-car space; today, it has 160,000 workers, including 15,000 engineers. "We are working day and night to develop this new technology," says Wang.

Wang is ambitious: he says that he still believes it is "reasonable" that BYD can be the largest automaker in the world by 2025. "China will be the world's largest economy. It will be twice the size of the United States. It will be the biggest market in the world. Chinese companies will be the biggest in the world. As we have the core competence, good quality, and manufacturing experience, we will have a chance to be number one in the world."

These are brave, and perhaps unrealistic, sentiments—but that doesn't mean his vision can simply be dismissed. I have heard many Asian entrepreneurs make similar statements. This seems to be partly a way of motivating staff ("you are part of something big and important") and partly a way of keeping the support of backers in government, who are key to seeing that banks and others continue to provide funding for growth ("stick with us and we'll provide jobs and growth and pay taxes"). The entrepreneurs themselves, optimists all, likely believe their own statements, whatever doubts they might have in the middle of the night. And although most do not become globally significant players, some do. I remember my skepticism when, in 1989, top executives at Samsung Electronics told me that they expected to surpass Sony. That same year, probably buoyed by the success of the 1988 Seoul Olympics, Hyundai Motors executives told me that they planned to rival General Motors. Samsung has long outstripped Sony, and Hyundai Motors, now among the world's largest automakers, has come a lot closer to catching up with a humbled General Motors (whose shareholders were wiped out during the 2008–2009 financial crisis) than most of us would have thought possible. So BYD's ambitions, improbable though they might seem, are not a complete fantasy.

Having branched out into vehicles, BYD now makes almost everything in the car except the tires and the glass. Staff boast of the company's vertical integration and a product cycle that, according to company officials, is one third the time of its competitors' cycles. However, this is not a tidy, tightly focused company. It is an opportunistic firm whose seemingly scattershot investments reflect the convictions of its founder. Batteries to battery storage to electric cars has a certain thematic consistency. Manufacturing solar panels, another business BYD has recently ventured into, is harder to understand, given the brutally competitive nature of what is increasingly a commodity business. So it is little surprise that BYD has now scaled back its solar ambitions, However, the company continues to sell standalone renewable energy systems for households, with rooftop solar powering BYD batteries which in turn provide electricity for the stove, refrigerator, and even an electric car. All of this is in keeping with Wang's aim to produce clean energy, store it, and then use it in environmentally friendly ways.

One reflection of China's position as the world's largest auto market is the hour it takes to drive from the busy Hong Kong border crossing at Luohu to BYD's Pingshan headquarters, though the distance is fewer than twenty miles as the crow flies. Although much of the route is on expressways, Shenzhen's highway traffic is stop-and-go. The core of what is today a sprawling city of some fifteen million people was a fishing town with a population of twenty thousand when Deng Xiaoping launched his economic reforms in 1978. Shenzhen was one of four initial Special Economic Zones—regions where Deng first experimented with the market reforms that revolutionized the country. "When you open the door, some flies will get in," Deng conceded about the Special Economic Zones and the less palatable parts of capitalism that they brought, but he convinced more conservative Party cadres that opening up to fresh air and new ideas was worth the cost.[17] Thanks largely to its proximity to Hong Kong, the Shenzhen Special Economic Zone was spectacularly successful—more so, by far, than the other three zones. Shenzhen thus was the cradle of modern China's economic takeoff, though the reforms pioneered there have by now been embraced by the country as a whole. Even today, it is a wild, anything-goes place. Almost no one in Shenzhen was born in the city, and the immigrants it draws are all striving for a better life. Shenzhen is far from Beijing and the capital city's political control. Its officials have typically been pragmatic and very business-friendly. This makes the region particularly well suited to an innovator like BYD.

Driving into BYD's headquarters, the site first looks fairly indistinguishable from the many massive manufacturing facilities in the region. But setting it apart is the parking lot, fairly sizable by Chinese standards and filled with employees' cars—all of which appear to be made by BYD. Although China is by far the world's largest car market in terms of sales volume, most Chinese still don't own cars; this luxury is largely confined to upper-middle-class workers, the wealthy, and government officials.

Wang—now the chairman and president of BYD—has an earnest, impatient manner, one that underlies a sobering speech he gave me during my meeting with him at company headquarters. "China will face a serious energy security problem in the coming ten to twenty years as Chinese people become richer and rely on oil. This will be a threat to national security. We must change our energy use. All the countries in the world will face this issue." He warns that oil could run out in fifty years. "We need a long-term vision, plan, thinking. We need to create new energy to support sustainable development. It's simple. We have sustainable power—solar. Even if the earth disappears, we will have the sun. One percent of the desert in China could produce enough power for all the Chinese people.... As an entrepreneur we must develop new energy that will be sustainable over 100 years."

Wang, in short, is an engineering-driven entrepreneur with an expansive vision. His dream is for a strong, energy-independent China. "China's energy security issue might be a little different. Energy security is very crucial right now for Chinese. Over 57 percent of our oil is imported—over 60 percent seriously threatens the security of the country. If China wants to develop the auto industry there is only one way to go, electric vehicles. It will be one family, one car in ten years. In China that means 400 million cars. There are 100 million cars right now."

A company video underscores this message: China faces a serious energy security issue. Rising affluence and more cars will leave the country short of oil unless there is a radical change. There will be a human cost, adds the video, with climate change producing some 200 million environmental refugees worldwide by 2050. There is a way out: BYD points to James Watt's steam engine in 1784, which led to the first industrial revolution; the Siemens generator in 1866, which led to electrification and the second industrial revolution; and Tim Berners-Lee's invention of the World Wide Web in 1990, which led to the third industrial revolution. Clean energy, the video says, will be the fourth industrial revolution, and BYD will be at the center of it. This is Wang's vision.

What the video omits is that the environmental benefits of using electric cars are not yet guaranteed—especially in a country like China, which still has an extraordinary reliance on coal. Although electric vehicles emit fewer greenhouse gases while they're being driven, these environmental benefits can be erased if they are charged on a grid that relies heavily on fossil fuels. Moreover, the production process for electric-car batteries is often environmentally damaging.[18]

Despite this uncertainly, China has been quick to get behind the nascent electric-car industry.[19] Government support, ranging from direct research and development subsidies to consumer rebates to mandated purchases of locally made electric vehicles, has made consumers more willing to consider electric vehicles. BYD has benefited from generous subsidies, from 2009 through 2013, totaling ¥2.3 billion ($335 million).[20]

In the face of consumer resistance to electric cars, BYD has repositioned itself to target the public transportation market. It's seen some modest success internationally—a London chauffeur company bought twenty of its battery-powered e6 cars, and Amsterdam's Airport Schiphol bought thirty-five of its buses to ferry passengers between the terminal and airplanes. From the Los Angeles airport to São Paulo to Warsaw to Montreal, small numbers of its buses are operating. Although BYD has had to deal with China's domestic protectionism—there are still more than one hundred vehicle manufacturers and officials favor local companies—this favoritism may be waning as the costs of pollution rise. One positive sign for BYD was the order in May 2014 from the city of Hangzhou for three thousand electric taxis and buses. Its latest electric car, the Qin, registered strong initial sales in 2014. BYD also has high hopes for the Denza, developed with Germany's Daimler and introduced in August 2014.

BYD shows much of what is best about the new breed of private-sector Chinese companies. Wang Chuanfu's entrepreneurial drive is backed by impressive engineering expertise and plentiful capital from private investors like Warren Buffett as well as by substantial support from the government. But its difficulties also highlight a number of key issues. These

A BYD electric taxi on the streets of Hong Kong. There were over six hundred in operation in Hong Kong as of April 2014. Photo credit: David McIntyre

include the struggle new manufacturers face in entering the demanding and competitive automobile market, the challenge many rapidly expanding Chinese companies have both in maintaining focus and in being able to deliver high-quality products in a timely fashion, and the slowness with which new technologies are accepted by consumers. That has left BYD dependent on subsidies while at the same time being disadvantaged by local protectionism in those areas where it does not have an assembly plant.

~

Electric autos are captivating. But they have yet to show that they can win broad consumer acceptance. Even in China, where a strong state has put forth a variety of pro-electric-vehicle policies and backed them up with significant funding for research as well as consumer incentives, the acceptance of electric vehicles has fallen far short of plan.

Toyota thinks that the answer lies not in pure electric cars but in hybrid ones.[21] Back in the 1990s, concerned about the cost and availability of fossil fuels as well as the environmental impact of CO_2 emissions,

Takeshi Uchiyamada demonstrates Toyota's Prius Plug-in Hybrid during a photo session in Tokyo in December 2009. Uchiyamada, dubbed the father of the Prius, has been chairman of Toyota since 2013. Photo credit: ©AP/Koji Sasahara

Toyota started developing a hybrid car that would have, in effect, two motors: a conventional gasoline engine and an electric-powered one. The goal was to be able to drive from Los Angeles to Washington, D.C., a distance of almost 2,700 miles, on one tank of gasoline. "It was a stretch idea," says Bernard O'Connor, the executive vice president of Toyota Motor Asia Pacific. "That was the dream. It was like going to Mars." Toyota introduced the first of the Prius hybrid line in Japan in December 1997, with its battery powered by energy that was recovered when the car braked and its gasoline engine supplementing the battery, especially when the vehicle was accelerating.

Prius sales started slowly; it took Toyota ten years, until 2007, to sell its first million hybrid vehicles. Since then, sales have accelerated dramatically, with more than 5 million units sold by the end of 2013. In 2012 alone, 1.2 million of the 8.6 million vehicles that Toyota sold were hybrids. Hybrids have gone well beyond the Prius; by late 2013, Toyota offered hybrid models for thirteen of its different brands, with plans on the way for fifteen more by 2015. So confident is Toyota of the increasing success of hybrids that

O'Connor expects cumulative sales of more than 10 million units by the end of 2015.

Toyota will keep broadening its hybrid offerings. In 2015, the company plans to introduce a hydrogen fuel-cell hybrid, replacing the battery with technology more suitable for heavier, long-distance vehicles such as trucks and buses. "We think it is too early for the [pure] electric vehicle to be a mainstream vehicle," says O'Connor, adding that the company's forecasts call for gasoline-powered (including hybrid) vehicles to make up 70 percent of the world's vehicle population well into the 2030s.

Jamshyd Godrej, head of India's Godrej & Boyce and a member of Toyota's international advisory board, points to the difficulty of creating an entirely new infrastructure that would allow consumers to recharge their cars quickly and far from home. Hybrid vehicles do away with this problem. "Toyota is able to piggyback on the existing infrastructure as well as banish range anxiety"—the fear many consumers have that their electric car's battery will run out of power and leave them stranded. Like landing a man on Mars, the dream of a hybrid that can go from Los Angeles to Washington, D.C., on a single tank of gas is unrealized—the Prius's range is a little more than one fifth of the distance. Although that part of the dream remains unfulfilled, the Toyota hybrid is an unquestioned commercial success. "It is the first vehicle to provide a serious alternative to the internal combustion engine since the Stanley Steamer ran out of steam in 1924," opined *Fortune* in 2006. "It has become an automotive landmark: a car for the future, designed for a world of scarce oil and surplus greenhouse gases."

~

Autos are not the only answer. The biggest, most easily exploitable transportation opportunity is in mass transport, especially in rail. Rail allows the density of Asian populations to become a strength and can move far more people at less cost than private autos or even buses, although requiring a smaller physical footprint. Rail's energy efficiency is magnified by Asia's push to build high-rise shopping centers, offices, and apartments near subway and train stations. Rail reduces pollution by getting passengers out of cars and buses. It reduces traffic congestion, further cutting pollution and thus improving fuel efficiency of the remaining vehicles on the road. It also improves overall economic efficiency by allowing for more concentrated land use, building high-rises, rather than encouraging low-rise horizontal urban and suburban sprawl. Asia is home to large-scale building and increased innovation in the field of rail-based transportation. Asia's railways, both urban mass transit and high-speed trains, already are among the world's leaders.

Japan's Shinkansen, the bullet train, was the world's first high-speed train. Its inauguration was timed to coincide with the 1964 Tokyo Olympics and exemplified Japan's rise after World War II. With trains traveling at speeds of up to two hundred miles per hour and an annual ridership of some 300 million passengers over nearly 1,500 miles of track, the Shinkansen system retains its place as one of the world's most successful rail systems. More recently, Korea, in 2004, and Taiwan, in 2007, have inaugurated high-speed trains. In all three of these countries, the train networks have changed overall travel patterns and contributed to economic and environmental efficiency by, in effect, extending the capital city to include more of the country.

The world's largest subway systems are in Asia, with New York's annual ridership of 1.7 billion trailing those of Tokyo, Seoul, Beijing, Shanghai, and Guangzhou. Almost every major capital city in Asia has a subway or light rail system or is building one. Kuala Lumpur, capital of middle-income Malaysia, has a high-speed train that takes passengers the fifty-five kilometers from the airport to downtown in twenty-eight minutes, hitting speeds of nearly one hundred miles an hour and all the while providing free high-speed wi-fi connectivity. Jakarta, a metropolitan area of over twenty-eight million people, is building its first subway, a $1.5 billion project backed by Japanese soft loans. Singapore is expected to double the length of its urban rail line to 360 kilometers by 2030, with researchers at Malaysian investment bank CIMB expecting ridership to grow from 688 million trips in 2012 to 1.3 billion in 2030.[22] Bangkok's elevated Skytrain has helped ease the city's chronic traffic jams and has pushed up the property values of apartments and offices that are close to a train station.[23]

China, though, is the most dramatic example of a country building a vast, fast rail network. China's high-speed train network is to that country what the interstate highway program of the 1950s and 1960s was to the postwar United States—an infrastructure dream that is transforming a nation. As recently as 1993, rail speed in China averaged only thirty miles an hour. Now, China is building the world's largest and most ambitious high-speed rail system. It has completed most of a network, which is expected to total 18,000 kilometers by 2015, at a cost of nearly $1 trillion. Already, its high-speed ridership is the world's highest, with 1.33 million daily passengers in 2012.[24]

As with many aspects of China's hyperpaced development, the high-speed rail system has had both safety and financial problems. In 2011, two trains collided, killing forty people. That crash led to revelations of massive corruption, undermining public confidence and literally slowing the program. Maximum train speeds were cut to 300 kilometers an hour from 350 kilometers, ridership dropped, and investment in new lines slowed. Finances are also a concern, with some Chinese experts warning that debt incurred for high-speed rail, in late 2012 totaling $640 billion, could trigger a subprime-style debt crisis, particularly if local resistance to the high-speed trains' relatively high ticket prices leads to lower ridership levels.[25]

Whatever its problems, the Chinese high-speed rail network shows the country's determination to get the technology needed to develop in a way that promotes energy security and lessens environmental damage, rather than relying on cars and roads, as large countries like the United States and Russia have done.

But in China, as in so much of Asia, rail is fighting a battle against time and against the relative ease with which new roads can be constructed. Rail requires significantly more capital investment and a longer construction period—in other words, a longer political horizon. Roads are expedient and provide some temporary relief. However, the tide may be shifting in favor of more rail. The experiences of Japan, Taiwan, and Korea—and now China—have demonstrated the social and economic benefits of high-speed trains, which knit countries together in a way that air and road travel cannot. Another positive example comes from Bangkok, where a metropolitan subway system built in the 1990s has not only had a positive impact on traffic but also rewoven the urban fabric, bolstering retail and commercial areas near Skytrain stations. That cities like Jakarta and Manila, whose infrastructure has badly lagged development, are now considering railway expansions is a testament to these successes (a final example of which is given below) and reflects a growing sense that the priority given to road development and private cars is simply not sustainable for the next few decades.

~

Hong Kong offers one of the world's great rail success stories. It is a story not of high-speed trains but of a company that is among the ten largest urban rail operators in the world and that also exemplifies environmental and financial sustainability. Hong Kong's MTR Corp., formerly known as

Hong Kong's MTR is one of the world's most efficient public transit systems.
Photo credit: David McIntyre

Mass Transit Rail, regularly appears at or near the top of global sustainability rankings. The MTR has lessons for cities around the world on how a concern with environmental sustainability can lead to better-run businesses and better experiences for customers.[26]

From its base in Hong Kong, the company currently manages urban rail networks in China as well as some as far afield as London, Melbourne, and Stockholm. The MTR takes a very long-term approach to its business: it is the world's first railway company to analyze the total carbon emissions embodied in its trains and equipment over the one-hundred-year estimated life of the track and fifty-year life of the cars.[27]

The MTR is an odd hybrid, for at its heart it is a property developer. Part of the MTR's strength comes from the fact that its home is in one of the world's most property-crazed cities, one where parents are more likely to spend weekends taking their children apartment viewing than to a sports match. The city has some of the highest property prices in the world—outstripping New York City's and behind only London—but it also suffers from roller-coaster swings in valuations. Prices fell 70 percent from a 1997 peak to a 2003 low. In the decade after 2003, average prices quadrupled, and in some cases, values rose five times or more. So it is apt that the Hong Kong government figured out how to harness the dynamism and the profitability of the property sector to fund mass transit.

Hong Kong's MTR is one of the world's busiest and biggest urban subway networks. Every weekday it carries more than 5 million passengers (average fare: 91 cents) in a city of some 7 million people. Its Tsuen Wan line alone moves 90,000 people an hour under the harbor between the Central district on Hong Kong Island and its grittier cross-harbor counterpart, Kowloon. Graffiti? None. Its stations double as underground shopping malls, so clean that even Tokyo's famed subway looks shabby in comparison. Yet it made a 2013 profit of $1.1 billion on revenues of $5.0 billion.

The MTR performs this alchemy by turning land into gold. In exchange for building rail lines, the Hong Kong government grants the corporation development rights at the stations. The company is one of the city's biggest landlords, building and then usually selling towering apartment blocks near the stations. The apartment residents ensure that there is plenty of traffic for the MTR's rail operations. The company profits further by letting out retail space at the stations. As of year-end 2013, it leased almost 267,000 square meters of commercial space that it owns—about one third more than the floor space in the Empire State Building. It acts as a property manager for an additional 90,000 residential units in Hong Kong, and its managed commercial and office space totals 763,000 square meters—an area slightly larger than the floor space of the World Trade Center twin towers.[28]

Hong Kong has some peculiarities that make this work. As a former British colony—and now as a semiautonomous Special Administrative Region in China—it had little choice but to grow vertically rather than horizontally, up rather than out. A far-reaching decision to set aside about one third of the city for parkland, later expanded to be more than 40 percent, further limited available space. Hong Kong is about the same size as New York City, and its population is about 7 million, compared to New York's 8 million; this parkland restriction means that an area equivalent to Manhattan, the Bronx, Brooklyn, and a bit more than half of Staten Island is ruled off-limits for development. This has vastly increased the effective density and pushed the city further in the direction of mass transit. Moreover, the government owns virtually all the land, generally leasing it for fifty years, giving the state an unusual degree of control.

In keeping with the small-government ethos that British colonial authorities hewed to during the 1970s and 1980s, the government initially put only a small amount of money into the system. In total, the government has invested a bit more than $4 billion in the past forty years but has made a gain (much of it paper profits in the form of stock) of more than $25 billion. The MTR currently has 218.2 kilometers of track and is in the midst of a $21 billion program that includes five major new projects. The showpiece of the current building program is a $1.3 billion terminal for China's high-speed rail network that, when it is completed in 2017, will make it possible to take a train from Hong Kong to Beijing in about eight hours, compared with some twenty-four hours today.[29]

The MTR more or less stumbled its way into becoming a sustainability leader. In the early 1990s, it hired Glenn Frommer to do its first environmental impact report. Frommer, a former rocket scientist who grew up in the Bronx, stayed on at the company and was pulled into the due diligence process when Goldman Sachs was readying the company for an initial public stock offering in the late 1990s. Goldman pressured the company into pulling together its various environmental risk factors and promising investors an annual sustainability report. Jack So, then the MTR's CEO, didn't see any objection to providing data on its various environment-related actions. "You can't argue with motherhood and apple pie," Frommer remembers him saying.

Spouting motherhood and apple pie–style bromides is one thing, but implementing a thoroughgoing sustainability program is far more difficult. The MTR is an engineering-driven company, and Frommer and his team took the mission seriously. As the company did the internal work to prepare its annual sustainability report, it began to understand its business better, and its sustainability efforts took on a self-perpetuating momentum. That allowed it to make investments that would pay back in future savings. As a global-scale transport company and a significant purchaser of new trains, track, and signaling equipment, it had heft with its suppliers. It worked with train maker Bombardier to develop trains that would use the energy in braking to generate power. It worked with lighting companies, air conditioner manufacturers, and escalator makers to figure out how to cut electricity use. From 2007 until 2011, the company cut its electricity use 15 percent per mile. It now is nearing the end of a program that will see it replace fluorescent lights with LEDs on many of its Hong Kong trains—a program that is projected to save more than 3 GW of electricity and 2,100 tons of carbon emissions annually after completion in 2014.[30]

With its buildings and trains, the MTR is among global leaders in virtually every measure of sustainability, from environmental factors to engineering, from health to business continuity readiness in the event of a crisis.

It is a member of the two leading stock market sustainability rankings, the FTSE4Good and the Dow Jones Sustainability Indexes. Its buildings conform to Hong Kong's highest energy-efficiency standards. It is an active member of the Carbon Disclosure Project and reports under the GRI. In 2012, it signed the World Business Council on Sustainable Development's Manifesto for Energy Efficiency in Buildings. That same year it adopted the Stock Exchange of Hong Kong's newly issued best-practice reporting code for environmental, social, and governance issues.

Climate change is a business threat throughout much of coastal East Asia, an area that has always been vulnerable to typhoons. The intensity of Hurricane Sandy, which caused so much damage to New York and New Jersey in November 2012, was roughly equivalent to what is known as a Signal 10 Typhoon in Hong Kong—the sort of storm that hits the city every decade or so. Concern about the impact of climate change has prompted the MTR to plan for much worse. It has adopted a sophisticated risk management system to take account of more severe and more erratic weather.

The MTR worries that a supertyphoon in connection with climate change could cause unprecedented flooding and damage. In 2012, Hong Kong raised the Signal 10 alert for Vicente, a storm that passed nearby and packed ten-minute sustained winds of ninety miles per hour. Eight of the thirty-five storms that hit the western Pacific region in 2012 were even more powerful than Vicente, with two storms seeing sustained winds of more than 125 miles per hour. Megi ("catfish" in Korean), a superstorm that passed south of the city in 2010, had one-minute sustained winds of 295 kilometers (183 miles) per hour and ten-minute sustained winds of 230 kilometers (143 miles) per hour. The MTR's trains are not designed to stay on the tracks in winds of more than 260 kilometers (162 miles) per hour. Other climate-change-related issues that the MTR considers include rail buckling as a result of heat, additional flooding as a result of more severe and more frequent storms, and the increased deterioration of infrastructure due to more erratic and intense weather.

"We've published risk assessments of climate change for the company, looking at potential threats and risks, including the higher cost of electricity and water and the impact of higher sea levels and more severe storms," says Frommer. "Each is assessed for the likelihood and consequence. We report this annually. I don't know how many companies do that. It certainly gives our insurers and our reinsurers much more confidence that we have an understanding of our business. That does affect our insurance premiums." For the MTR, climate change in particular and environmental concerns in general are all about understanding and running its business better. An element of this involves the company's social license to operate, given the need to win support for activities—from construction to fare increases—from broad sections of the population. At least as important is the more practical impact of using resources ranging from water to electricity more efficiently.

The focus on sustainability has led the MTR to build additional resilience into its operations. That costs money, but Frommer says that it is money well spent—among other things, it leads to lower insurance premiums, as the company is able to show its insurers that it has an increased understanding of its business and is thus less likely to suffer losses.

~

Although the traditional internal-combustion-powered automobile is still king in Asia, new ideas could threaten its supremacy. Toyota's pioneering hybrid technology, which does not require any changes to the existing infrastructure of gasoline stations and is an easy transition for drivers, is particularly promising. Electric vehicles will find a niche, but the experience of the past decade shows the difficulty that the new technology has had in establishing itself as a mainstream alternative. Given Asia's increasingly urban population, filled with people who increasingly have the money and the need to travel, rail offers a compelling alternative. Nothing will replace the freedom and the possibilities offered by the individual car, but it should be complemented by extensive and pleasant public transportation systems of the sort that Japan has if Asia is to avoid the prospect of the weeks-long, 1,000-plus-mile traffic jams that China periodically experiences. Rail could and should be to twenty-first-century Asia what roads were to the twentieth, a way of promoting economic prosperity and drawing together cities, nations, and their people.

Throughout Asia's transportation industry, this is a time of sometimes radical experimentation, from the modest Reva electric cars to BYD's electric buses, from Toyota's hybrids to Hong Kong's MTR and China's high-speed rail network. Asia has the strong economic growth and the corporate and government financial resources to reshape transport networks and the cities and countries that they serve. In transport, as in cities and buildings, the best Asian governments and companies are looking for alternatives to the region's pressing resource constraints. Some will succeed, and others will not, but Asia's financial resources will ensure that this process of wide-ranging experimentation continues at a rapid pace, one unlikely to be matched anywhere else in the world.

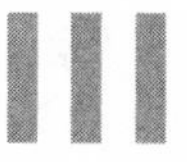

Nature: *Forests, Farms, and Water*

> You can't just demolish everything.
> —JAKARTA GOVERNOR JOKO WIDODO

Indonesia's President Susilo Bambang Yudhoyono is dressed informally—slacks and an open-necked shirt—and pointing off-camera. Beside him is Foreign Minister Marty Natalegawa, more formally dressed—dark suit, white shirt, and striped tie—and wearing a slightly bewildered expression.

The foreign minister's strained look is understandable. For all their seeming normality from the waist up, in this extraordinary photo the then-president and the foreign minister of the world's fourth most populous country are standing in calf-deep water, their trousers rolled up past their knees, in the presidential palace in downtown Jakarta. Floodwaters had rolled through the city, and on the streets outside the palace, cars were stranded. More than two dozen people were killed in these January 2013 floods after a mere one hundred millimeters, or four inches, of rain fell on the city.

This sort of occurrence is all too frequent in Jakarta and many other Asian cities. In 2007, guests had to be ferried out of one of Jakarta's posh Four Seasons hotels in plastic laundry tubs, part of an evacuation effort that affected more than 300,000 people.[1] As metropolitan Jakarta, home to more than twenty-eight million people, destroys mangrove swamps and other water absorbers, flooding will be more and more frequent. After the 2013 floods, Jakarta Governor Joko Widodo (who in October 2014 succeeded

Indonesian President Susilo Bambang Yudhoyono (second from left) and Indonesian Foreign Minister Marty Natalegawa (left) inspect the flooded Presidential Palace in Jakarta, Indonesia, on January 17, 2013. Photo credit: © Xinhua/Dudi Anung/Corbis

Yudhoyono as president) noted ruefully: "The Dutch built 300 dams and lakes, but there are only 50 left. The wetlands, woods and other green spaces north of the city have been taken over by housing complexes and malls. You can't just demolish everything."[2]

The environmental challenges posed in the wider natural world are some of the most difficult. First, by definition, these areas are the largest. They encompass forests, farms and plantations, rivers, and oceans. Loggers, farmers, and fishermen—the people who exploit forests and oceans for their living—are literally living on the margin of the human world. Many—and in Asia certainly most—of the people who work the fields, forests, and oceans are poor, more concerned with scraping out a daily living than with worrying about climate change or biodiversity or sustainable development.

Water is a particularly challenging issue. Water covers three quarters of the planet and makes our Earth unique in our solar system. It is a basic need, something we require for survival; none of us can live more than a few days without water.

Still, we often treat our water sources as a sewer. The situation in much of Asia is alarming. The Asian Development Bank says pollution levels in Asia's rivers are four times the global average and twenty times levels set by the Organisation for Economic Co-operation and Development. Fecal waste averages fifty times World Health Organization guidelines. Water shortages are likely to be exacerbated by climate change, especially in China and India, as drought follows the melting of Himalayan glaciers.

Water is mostly used in agriculture. There is not very much good news. National policies have rarely treated water as a scarce commodity. In India, free electricity supplied to farmers allows them to run ever-deeper pumps that deplete water sources. It is left to forward-looking parties to fight this trend—for instance, Esquel, a Hong Kong–based textile company, has helped cotton farmers in the arid western Chinese province of Xinjiang finance the installation of water-saving drip irrigation systems in an effort to mitigate the impact of cotton growing, the most water-intensive part of its shirt manufacturing business. Esquel and other companies and governments that are encouraging innovation in water use are the focus of chapter 6.

Consumers can play an important role in sparking change. This is a significant difference between agricultural and marine products, especially food and other consumer goods, and the areas of energy and the urban world discussed in the first two parts of this book. Consumers cannot stop China from building more coal plants and have at best an indirect effect in promoting green buildings and more energy-efficient cities. Even the relative popularity of fuel-efficient vehicles and gas guzzlers to date owes as much to government policy and the relative availability of subsidies and charging stations as to consumer choice. But consumers can have a significant impact when it comes to natural resources like water, forests, and the biodiversity they promote.

The importance of Western consumers was dramatically shown in the reaction to a 2010 Greenpeace spoof advertisement savaging Nestlé for its contribution to tropical rain forest deforestation. That attack ad, discussed at greater length in chapter 7, prompted an overhaul of Nestlé's palm-oil procurement policies. In timber, too, major wood buyers are insisting that Asian suppliers hew to Western standards. Ikea's furniture-buying strictures are helping to reform forestry practices from Borneo to Siberia. Indonesian forestry giant Sinar Mas in early 2013 announced sweeping changes to its forestry practices in response to supplier pressure. Palm-trading colossus Wilmar did the same later that year, for the same reason. Whether it is to stop the felling of old-growth forest for logs used to make tissue paper or furniture or the purchase of agricultural commodities such as palm oil and coffee that are being grown using environmentally unsustainable methods, the most effective way to effect change has been through image-conscious multinationals.

Forests, farms, and water: both figuratively and literally, these areas comprise a vast amount of territory. The following two chapters, comprising the last of the book's three main parts, examine the major challenges, the opportunities for change, and some of the companies that are doing the most innovative and promising sustainability work.

6

"Water Is More Important than Oil"

> Water scarcity is set to become Asia's defining crisis by midcentury.
>
> —BRAHMA CHELLANEY

Pigs, floating, dead, by the thousands—more than ten thousand of them were scooped out of the Huangpu River near Shanghai in early 2013. No one seemed to know if they were killed by the cold or a swine flu or why they were dumped rather than being properly disposed of. But the pictures of a mysterious flotilla of dead animals in a river running through China's major commercial city summed up the state of Asia's waterways.

The response of China's citizens, weary of polluted air, tainted food, and cancer villages, was one of sardonic resignation.[1] Although protests against toxic water spills and the cover-ups that typically follow are increasingly common in China, there were few protests in response to this strange event; perhaps the absurdity of the floating pigs made protests seem futile, for humor was a common response. A joke made the rounds: Beijingers, always proud of the capital, boast that the city's thick smog meant that they only had to open their windows "and we get free cigarettes." That's nothing, scoffed the Shanghainese. "We turn on the water faucet and we get pork soup!"[2]

Floating pigs are not common, but water problems are all too frequent. Floods in Asian cities are increasing from Shanghai to Jeddah. Manila, Bangkok, Beijing, and Hanoi are among the Asian capitals that have had serious floods in recent years. Badly planned urbanization in low-lying cities, oceans that are rising, and a climate that is producing more extreme weather have all contributed to more water-related disasters.

Yet Asia's bigger problem is not too much water but not enough. A combination of relatively little water per capita in much of Asia, an unprecedented number of people moving to cities, and the mismanagement of water resources adds up to a complex set of interrelated problems that get to the heart of daily life for many Asians. Asia's per capita availability of freshwater is half the world average. China and India together make up almost 40 percent of the world's population but have only 10.8 percent of the water.[3]

Places without water do not control their existence. One reason the British could not keep Hong Kong after 1997, when a ninety-nine-year lease ran out on the colony's New Territories, was that Chinese control over the city's water meant that to defy the mainland was unthinkable. Singapore (discussed below) has worked assiduously in the nearly fifty years since independence to avoid being placed in a similar situation vis-à-vis its neighbors.

Asia's population has roughly tripled in the past six decades, from 1.4 billion in 1950 to about 4.3 billion today, and this hugely expanded population is putting enormous pressure on water resources. Most Asians do not have the sort of secure access to water that people in the United States, Europe, Canada, or even drought-ravaged Australia take for granted. In China, in a telling indication of how rare access to clean water is, the Tianjin Eco-city boasts as one of its unique selling points the promise that residents will be able to drink water straight from the tap. Water in much of China is not only polluted but also increasingly scarce. Northern China has less water per person than Saudi Arabia. On the other side of Asia, Sultan Ahmed Al Jaber, the CEO of the ambitious new eco-city city of Masdar in Abu Dhabi, says that water "is more important than oil."[4]

In his authoritative book on water in Asia, Brahma Chellaney, a professor at New Delhi's Center for Policy Research, contends that Asia's future economic prosperity is at risk because of water issues. The continuing success of the green revolution, the transformation that saved something on the order of one billion people from starvation as a result of a dramatic improvement in agricultural yields, is in doubt. He is particularly critical of China, where many of Asia's most important rivers originate, for its unilateral policy of damming and diverting the upstream reaches of rivers including the Mekong. China, he says, "has made the control and manipulation of natural river flows a fulcrum of its power and economic progress, spurring intensifying concern among downstream nations."[5] Indeed, Chellaney warns of war, or at least conflict, as a result of water shortage.

Despite the low availability of water, Asia remains extraordinarily wasteful in its water use, and government policies often encourage this waste. As with so many other resources, improper pricing has contributed to unsustainable practices. Asians use relatively more water for agriculture than is the global norm: in Asia, more than 80 percent of water goes to agriculture, compared with a global average of 71 percent. Rice is a very water-intensive crop, needing 1,550 to 2,000 liters of water on average to grow a kilogram, but irrigation in Asia is inefficient. "Most Asian farmers lack financial resources and knowledge to invest in more expensive water-saving technologies, like drip irrigation infrastructure or storage," says Chellaney, yet state subsidies have helped weaken "price signals, tempting farmers to take too much water from rivers, over-pump groundwater and generally waste freshwater resources." Cheap or free electricity, combined with inexpensive pumps, allows the depletion of underground aquifers.

However, according to Chellaney, the fastest increase in water withdrawals in Asia since the 1990s has come from urban households and the industrial sector, signs of the rapid urbanization and industrialization taking place across the continent. Coal and nuclear power plants are big users of water, and the demand for energy keeps multiplying. China has an aggressive biofuels target, calling for twelve million metric tons by 2020; Chellaney quotes one study saying that this would require 5 to 10 percent of all the cultivated land in the country, as well as thirty-two to seventy-two cubic kilometers of water a year, roughly equivalent to the annual discharge of the Yellow River.

A handful of companies are treating water as a scarcer commodity than governments do. Hong Kong–based Swire, for instance, is implementing a groupwide sustainability reporting system that includes water as a key focus. It claims that its Coca-Cola bottling plant in China's Henan Province already is among the most energy- and water-efficient bottling plants in the world. "The aim is for all plants to return all the water which they use to make beverages back to the water supply system by 2020," says Philippe Lacamp, head of the sustainability unit at John Swire & Sons.

Asian governments, for the most part, have been slow to rise to the challenge posed by the looming water threat. However, change is coming. A number of protests in China in recent years against unsafe drinking water—usually as a result of toxic spills—have unnerved authorities.[6] Using policy tools ranging from higher prices for water to privatization of state utilities to desalination and water-efficiency efforts to massive engineering projects, governments are scrambling to ensure that supplies of clean water can keep pace with economic and social development.

This chapter highlights three very different cases that show the actions some governments and companies are taking in the face of water scarcity, deforestation, and other environmental challenges. This first is the Ayala Group, one of the Philippines's largest conglomerates, which has stepped in to provide water where a state-owned water company had failed. Building on its success, Ayala's Manila Water subsidiary is partnering with the Philippine government and the Asian Development Bank to design a broader water-management strategy that seeks to protect watersheds, clean up the Pasig River running through central Manila, and develop more resiliency to cope with the devastating typhoons that are likely to continue worsening as a result of climate change. The second case is Singapore, a nation whose very existence was at risk because of lack of water. Today, about one third of its water comes from treated wastewater. This is a case where a national strategic imperative has led to far-reaching innovation. Singapore has emerged as

NEWater's visitor center in Singapore. Photo credit: David McIntyre

a global leader in water treatment technology and has spawned a number of ambitious, internationally minded water companies. The third case is from the manufacturing sector and looks at how one of the world's largest shirt makers, Hong Kong–based Esquel, has embraced water conservation and treatment as part of a hard-headed business approach that has led to both increased innovation and cost savings. Each case provides valuable lessons for a future in which humans will be more plentiful—and water less so.

~

We stepped out of our van and walked a few hundred meters into the Santa Elena barangay, a modest village tucked among trees in the midst of Marikina City, part of sprawling metropolitan Manila.[7] We were surrounded by expensive gated housing estates, but our destination was one of the humble informal settlements that dot Manila and so many other fast-growing Asian cities. People here are on the slippery lower edge of middle class. In this country, with its high density and poor infrastructure, they are especially vulnerable to the typhoons that regularly sweep through the country. The Philippines lies in Asia's typhoon belt, bearing the brunt of more hurricane-force storms than almost any major nation on earth. Metropolitan Manila gets hit by an average of three to five typhoons a year. Typhoon Haiyan, the country's deadliest storm in modern times, killed at least 6,200 people and caused more than $800 million in damage in November 2013.[8] With a population of close to one hundred million people—more than Germany's—the Philippines has the climatological bad luck to combine a surfeit of water with recurring droughts.

About twenty million Filipinos are concentrated in and around the capital of Manila, many of them in barangays—the smallest of the country's administrative divisions, something akin to a city ward or a rural village—like Marikina City's Santa Elena. Irene Abella, a high school mathematics teacher and a part-time official, had been living there for sixteen years when I visited in August 2010. Abella, forty-two years old with two sons, moved to Santa Elena from an outlying province, as many of her neighbors did.

Abella tells me that her family usually pays fewer than 200 pesos ($5) a month for its water. Before the water system was privatized and Manila Water's service came to her area—she isn't too sure, but she thinks she got running water in 2002 or 2003—the family paid 1,000 pesos. Worse than the cost was the hassle. It could take Abella half a day, two times a week, to wait in line to fill her big containers of water and lug them home.[9]

Gerardo C. Ablaza, Jr., Manila Water's president and CEO (seated in the middle), poses for a photo with his management team and the Pasig City government's Green Police volunteers to highlight the revival of the Pasig River. Photo credit: Manila Water

That Manila Water has managed to serve millions of customers while cutting prices is impressive, given the challenges associated with operating in the city. A major one is population growth; metropolitan Manila's population has grown almost tenfold, from just over 1.5 million in 1950 to an estimated 22.7 million in 2014—almost twice the combined population of Hong Kong and Singapore. And although in the 1950s the Philippines was one of Asia's most prosperous and promising countries (the World Bank lauded it as a nation poised for an economic takeoff), poor governance has stifled economic growth. Services like electricity, telephones, and water have lagged.

It was only in the 1990s that a series of reforms under President Fidel Ramos started to resolve the worst of the bottlenecks. In electricity, private operators were allowed to compete against a state-run utility, and the country's chronic brownouts were largely eliminated.[10] Mobile phone operators allowed consumers to sidestep what had often been a frustrating, years-long wait for a fixed-phone line. Other reforms took place in banking and shipping—and in water. In the mid-1990s, the government undertook the world's largest water privatization, splitting the capital into two zones and selling Manila's water services to two groups of private investors. The newly formed Manila Water took over the eastern part of the capital's service area at the beginning of 1997. At the time, only two of three people in the service area had access to tap water. Now 99 percent do.[11]

Manila Water's secret to success was fairly simple. It extended the system to more households, while dramatically reducing the amount of water lost to waste and theft. In 1997, before Manila Water took over from the government, two thirds of water supplied by the municipal water company leaked or was stolen. In 2013, it was 12 percent, a figure that compares favorably with London (25 percent) and most other major cities. The company invested the additional revenues in expanding its network of piped water, getting even more customers in an area with a rapidly growing population. More people paying for water means that more money is available for investment in water pipes, treatment facilities, and general watershed management. It also means that in drought years more people will have access to water.[12]

Manila Water serves more than 6 million residents in the capital, more than three times as many as it did when it took over the aging municipal waterworks in 1997. Although Abella saw her water bill cut to a fifth of its previous amount, the International Finance Corp. estimates that average savings have been even more dramatic—many of the houses that Manila Water now serves pay one twentieth what they did under the public waterworks. "There is a very different discipline, one of execution, in the private sector," says Manila Water President Gerardo (Gerry) Ablaza. "This is what we are built to do. Government is not built to run utilities or business units. There is the perception that government's hands are tied in paying competitive market compensation for talent. We have the capacity of attracting very good, highly competent people." Still, more than fifteen years after the utility was taken over from the government, Ablaza proudly notes that 70 percent of Manila Water's employees remain from the state-owned predecessor.

Manila Water has been one of the world's great success stories in its ability to supply a basic good to city-dwellers, many of whom are poor. It has done this while slashing the price most households pay for water and helping realize national objectives related to cleaning up sewage-choked rivers and adapting to climate change. At the same time, Manila Water has given its shareholders a solid return, with its dividends and stock appreciation returning an average of 18 percent a year to investors from 2007 to 2011. In 2012 the return to shareholders jumped to 67 percent.

Manila Water has won numerous accolades, notably those from the World Bank's International Finance Corp., the bank's private-sector financing arm and an investor in Manila Water. The World Economic Forum, organizer of the annual Davos summit in Switzerland, named Manila Water one of the sixteen "most environmentally aware" companies in the developing world in 2011. Riding on its success, Manila Water has expanded into new service areas in other parts of the country, including the former U.S. Clark Air Base and the tourist island of Boracay, and has also looked abroad to Vietnam.

Given the political sensitivity of profiting from an essential resource, Manila Water's impressive financial returns necessitate a consistent focus on the benefits that customers and stakeholders receive. To try to win community backing, the company has worked to see that more people, including those on lower rungs of the economic ladder, have access to its water. When it started service, only two-thirds of households in its service area had water—and even most of those who were lucky enough to have water taps couldn't count on always-on, 24/7 service. That left many residents with no choice but to buy water from tanker trucks and cart or carry it hundreds of meters to their homes, often over unpaved or poorly paved paths. According to company calculations, it has doubled the number of people served, adding 3.2 million customers, more than half (1.7 million) from low-income communities. Its strategy to win grassroots support includes programs to improve access to water. The Tubig Para Sa Barangay (Water for the Barangay) program focuses on getting taps to poor households. When installing new taps, the company has taken a flexible approach to the lack of proper legal title for property, removing a major barrier to service. The World Bank provides funding for a sister program to subsidize connection fees for low-income residents. In all, the company says 99 percent of the residents in its metropolitan Manila service area have access to tap water.[13]

Manila Water's environmental and social programs build on an existing framework. The company is part of the Ayala Group, one of the country's largest and oldest conglomerates. Ayala has more than 175 years of history in the Philippines and regularly ranks among the nation's best-regarded business groups. Ayala has also taken an increasingly sophisticated approach to environmental, social, and governance issues and is widely considered one of the regional leaders in this regard. Ayala and in turn Manila Water try to align their work with broader government planning. The latter also says that it is working to help the country meet the Millennium Development Goals for access to good drinking water and proper sanitation.[14]

Manila Water has continued its parent company's focus on sustainability in a number of ways. After the company's impressive performance in its first decade of operations won it, in 2009, a fifteen-year extension on the original twenty-five-year concession agreement, it was able to fund a wide-ranging 450 billion peso (about $11 billion at current exchange rates) investment program. It's used this to help clean up polluted rivers and to provide employment through a variety of small-scale enterprises related to the core water business. In nearby Villa Beatriz, Barangay Matadang, the company has contracted with a local cooperative to make meter protectors (to prevent tampering) and other related equipment. It is a small venture, employing about a dozen people and generating sales of more than 1 million pesos a month. Established in 2005, the cooperative was originally housed in a modest one-story building. By the time I visited in 2010, it had rebuilt the site with a three-story building, one of the nicer ones in the area. Manila Water calculated that in 2011 it put 4.1 million pesos back into local communities through eight accredited cooperatives like that in Villa Beatriz, providing employment for more than one thousand people. These co-ops provide goods and services ranging from the meter protectors I saw to car washing and printing. Other small, local construction businesses have benefited from Manila Water's ongoing spending to lay new water pipes.[15]

Another 78 billion pesos of the fund are allotted to developing new water sources and connecting households to Manila Water's sewerage pipes, rather than forcing households to rely on septic tanks or dump untreated waste into rivers and watersheds. Eliminating sewage going into rivers will allow the cleanup of the San Juan, Marikina, and Pasig Rivers.[16]

Providing water and wastewater treatment is something that governments usually do; with Manila Water taking it on in the Philippines, people now have to pay for it. Untreated sewage is nominally free, although there is a cost to society in the form of sickness, disease, and a less pleasant living environment. Asking people to pay for sewage treatment as part of their water is politically difficult because it means that overall water bills are higher without any obvious immediate benefit. Treating waste is expensive: the costs to build and operate wastewater treatment plants are about three times those of water treatment plants, Ablaza says. "When people have gone for years without drinking water and then they have it coming out of the faucets 24/7 they say 'I will pay.'" But when it comes to sewage and wastewater being dumped, "95 percent of them don't know what is going on with the rivers and they don't care." Manila Water has set up Project Zero to try to make its wastewater treatment plants cost-neutral by converting waste to energy. The aim is to keep costs under control and benefit the environment.

Cleaning up sewage has economic benefits. The tropical island of Boracay, for instance, was literally choking on its own development. The island's septic tanks and sewage system could not keep up with growth, and the effluent runoff was killing the coral, coral that holds up the white sand for which Boracay is famed. A wastewater treatment plant built by Manila Water has improved the situation and encouraged tourists to return to the island. "We are very proud to say that we are contributing to the sustainability of tourism in Boracay," says Ablaza. The benefits go further that this; the spending on cleaning up water sources, cutting water losses, and treating wastewater will, according to Manila Water's 2009 Annual Report, help "to meet the changing circumstances brought about by climate change."[17]

Those circumstances seem to become more alarming every year. In 2009, the issue was drought, though Manila Water was able to avoid water rationing, thanks to its continuing efforts to reduce the amount of water lost to leakage and theft; that year the company achieved a new low "non-revenue" water rate of 16 percent, a figure that continued to drop in later years. In 2009, Typhoon Ketsana (Ondoy to locals) devastated Manila, with 450 millimeters of rain (almost 18 inches) in a single twelve-hour period, flooding much of the capital. It was one of the worst storms in decades, the sort of flood that statistically should occur only once every 180 years.[18] More than three hundred thousand people in metropolitan Manila were affected, and Manila Water saw eight of its nineteen facilities (mostly pumping stations) flooded, meaning that many people lost water for days. "These natural calamities awakened all of us to alarming social and environmental realities that are simply too serious to ignore," wrote the company's chairman, Fernando Zobel de Ayala, to shareholders in the wake of the storm. "They beg us to address the realities of poverty and environmental preservation. These are key imperatives today, particularly in the business sector, as we seek to ensure our society's long-term survival and the sustainable growth of our economy."[19]

In 2012, just three years after Typhoon Ondoy, four hundred millimeters (almost sixteen inches) of rain fell in Manila in just twenty-four hours as the Philippines was peripherally hit by Typhoon Haikui. "We thought these types of rains fell every fifty years," said Ablaza when we spoke in early 2013. "This is the new normal." Manila Water's mitigation efforts were far more successful in 2012's Haikui, with thirty-five of thirty-six facilities continuing to operate normally and water supplies unaffec ening "new normal" was underscored in late 2013 when lashed the southern Philippines. In the aftermath of Haiyan, was able to maintain uninterrupted water supply and wastewate tions in Boracay, the part of its service area that was hit hardest by t storm. It ensured the area's evacuation sites had drinking water and opened a water treatment plant to provide water to residents. It also set up mobile water treatment plants (a step it also took after an earthquake hit the island of Bohol in October 2013), which are able to produce two thousand liters of potable water an hour using any available water, outside of its franchise areas. As it has done with wastewater treatment, the company is taking on some of the disaster relief functions that in more developed countries would be undertaken by the government. "Disaster response is certainly another contribution we can make," says Ablaza. The resiliency and contingency planning developed to deal with the more severe weather brought about by climate change will help protect in the case of calamities, "whether they are storms or earthquakes."[20]

Manila Water is also looking at its carbon footprint, both direct and indirect, and is performing weekly monitoring of its electricity consumption.[21] Ablaza concedes that the company has lagged in this measure. "It is very difficult to make a substantial dent on our carbon footprint. It tends to increase with the more pumping stations we build. We are monitoring the amount of energy we consume per mcm [million cubic meters] of water. This is part of a weekly monitoring that top management receives." Ablaza says that the company is trying to move away from pumping water as it is produced to pumping it as supplies are needed, so as to ensure that water is not pumped unnecessarily.

In short, Manila Water recognizes that if it wants to succeed, it must do more than just deliver water. It must engage on a broad level with many aspects of society and the state—working to serve the poor, helping to create jobs, and coordinating with the government to develop and implement a national environmental strategy. The government's decision to extend the Manila East concession period by an additional fifteen years, to 2037, reflects the success Manila Water has had. That lengthened concession period in turn has given it the certainty to promise the 450 billion pesos of investment for river cleanup and job creation. In short, the company reflects the ethos of a business group that has been in the Philippines almost two centuries and understands that true business success requires a social license to operate.

The debate over the roles of the state and the private sector is an evolving, global one. Different countries will have different solutions. Manila Water is a private company, but in the Philippines, it is being asked to play a national role. It is asked to protect watershed lands, to build or rehabilitate dams, to clean up the Pasig River that runs through the capital, and to help provide national disaster response. This reflects the relatively stronger capacity of the private sector in the Philippines and the relative weakness of the government. Ablaza contends that Manila Water even helps relieve some of the nation's financial burden. The Philippines is a country that has long struggled with heavy foreign-denominated debts, and the company's ability to tap international capital markets lessens pressure on the country's sovereign credit ratings. "We have access to capital," says Ablaza. "Obviously, the government does, too, but it has very severe constraints, especially if it has to borrow."

The country's reliance on the corporate sector necessitates robust regulatory oversight to ensure that private companies like Manila Water are meeting measurable objectives in everything from customer service to watershed management. "The government has to be the architect of a master plan," says Ablaza. "We are only contractors. The government should set the vision for water—the sourcing of water, dams, reforestation. There has to be a holistic vision of water from end to end. Government should be at the forefront of making the public understand why the economic value of water needs to be reflected in its price."

For all the international praise Manila Water has garnered, it has had to contend with numerous domestic critics, some offended by the notion that a private company should profit from selling water and others unhappy with its impressive profitability. In 2013, the company battled attempts to redefine its tariff structure, prompting a coalition of business groups to come to its defense. The business groups attacked what they called "reckless statements (such as 'grossly unjust payments') and calls for unilateral changes—even abrogation of contracts—that have been publicly aired, including by government agents whose responsibility is to deliver on obligations under those contracts."[22] Whether the company has brought piped water into more households or lowered the bill for most of these people is beside the point for critics. Certainly, Manila Water has won numerous accolades at home and abroad. It has hewed to high standards of governance and transparency, with annual reporting complying with the rigorous Global Reporting Initiative G3 standards, which cover a range of human rights, social, environmental, labor, and other issues. That a company like this nonetheless faces criticism is a reminder that any providing as basic a human need as water must wor license to operate.

~

Singapore is a tropical island that gets massive amour —about 2.4 meters (almost 8 feet) a year.[23] Smaller than New York City, like New York, it lies on the ocean. There is little in the way of aquifers or natural groundwater, and its reservoirs cannot hold enough water for a city of five million people. Relations with neighbors Malaysia and Indonesia are often uneasy, so importing water makes Singapore vulnerable. Water, Singaporeans know, is an issue of national survival.

Singapore won its independence in 1963 as part of a Malaysian federation and then broke away from Malaysia in 1965. The two nations have long had an uneasy relationship, with Singapore's majority Chinese population and Malaysia's majority Muslim population eyeing one another warily. And water has long been at the heart of the two nations' relationship. Indeed, after gaining independence from Great Britain in 1963, Singapore remained in the larger Malaysian federation in part because of concern that the small island was not self-sufficient in terms of water. To this day, Singapore still relies on Malaysia for much of its water under a preexisting agreement signed during British rule.[24]

Singapore's founding father, Lee Kuan Yew, wrote in his memoirs that one reason the country developed a military was to guarantee water supplies. Although Malaysia never did cut off water, the threat was often just beneath the surface. In Malaysia in 1998, a crowd gathered for a speech by Prime Minister Mahathir Mohamad chanted, “Cut, cut, cut,” in reference to the water the country was supplying to Singapore. Singapore looked to its other neighbor, Indonesia, as a way of diversifying its dependence on Malaysia, but this was not a panacea—in 2000, Indonesian President Abdurrahman Wahid said water supplies to Singapore could easily be cut off by a “Malaysian-Indonesian alliance.”[25]

Given all these uncertainties, the Singapore government has made access to clean water a national priority. A wide-ranging strategy—the so-called Four National Taps program—attempts to balance water supplies among local reservoirs, imported water, desalination, and treated sewage water. “It is an issue of survival,” says Chew Men Leong, the CEO of Singapore's Public Utilities Board (PUB), who took over as head of the national agency after he stepped down as naval chief in 2011.[26] When leaving the Malaysian federation in 1965, says Chew, "we needed to figure out how we were going to survive long term. Lee Kuan Yew and his wife saw the necessity to entrench the water agreements [with Malaysia]. They put the water agreements back to back with the separation agreement and the documents were lodged with the UN."

Singapore is probably the only country with a water treaty stapled to the back of its declaration of independence. Lodging the two documents together at the United Nations internationalized the agreement and, says Chew, "tells you the importance. We realized water was a strategic weakness. To move forward as a country we were engrossed with the idea of how we got water security in the long term. Water is the policy that dominates almost every other policy that the government needs to come up with."

Most government-run water utilities count themselves lucky if they are able to recover their costs. In Singapore, by contrast, water is treated as a scarce good and priced according to the marginal cost of additional water sources. In other words, users are not paying for existing facilities but are instead charged what it would cost if the government had to bring on a new desalination plant or build the facilities to recycle additional water. "I don't think there is any other country in the world that does this," says Chew. This spending is significant. A massive $2.7 billion waterworks program now has centralized collection of used water, for recycling, on the eastern side of the island. A tunnel of a similar scale is slated to follow for the western side.

In the 1960s and 1970s, Singapore's strategic focus was to collect as much rainwater runoff as possible. The efforts have paid off. Today, a remarkable two thirds of the country serves as water catchment, feeding the island's reservoirs. Multibillion-dollar tunnel networks collect water from more urban areas, feeding it to treatment plants. The effort is globally unparalleled. "It is a huge experiment," says Chew.

Singapore has stronger technical capacity today and has expanded its focus to desalination and wastewater treatment. With water needs expected to double over the next half-century, Singapore officials believe that technology is the key to further innovation in water and have devoted significant funding and manpower to build the country into a self-styled "hydrohub." The government has committed S$470 million ($375 million) for the industry and hopes to double employment in it to around 11,000—mostly skilled and professional workers—by 2015. At the same time, it hopes that the value-added contribution to the economy will jump from about S$500 million in 2003 to S$1.7 billion in 2015. There are now twenty-five water-related research institutes and corporate laboratories in the city-state, with a total of S$221 million, or about $180 million, spent on 348 water-related projects.[27]

With this push of funding, Singapore hopes to develop itself as a research center on water issues. Some of the projects under way include combining DNA sequencing with reverse polymerase chain reaction to profile smelly and off-flavor bacteria in reservoirs so that outbreaks can be dealt with earlier; using radar data, rainfall forecasting, and detailed drainage information to forecast localized flooding on a real-time basis; and exploring variable salinity plant technology, which would allow Singapore to dramatically increase its water catchment efficiency.[28]

Government and academic institutions often partner to develop such technologies. One interagency body, the Environment and Water Industry Program Office (EWI), includes the powerful Economic Development Board and other agencies as well as two of the country's leading schools, the National University of Singapore and Nanyang Technological University. Although the EWI was set up only in 2006, it has helped put Singapore on the map for research and innovation in water. There are also links between its universities and peers around the world, with collaboration among the National University of Singapore and Oxford and Peking Universities as well as between Nanyang Institute of Technology and the University of New South Wales.[29]

There are also close alliances between academia and industry, with an emphasis on commercialization. The TechPioneer program provides funding of up to S$10 million for users to introduce new technologies into water or environmental processes. This allows for real-world testing of new processes. More than one hundred projects have gone through this test-bedding process since 2007. The government has also brought in private companies to build and operate some of its water facilities. "We bring the industry so that we can tap on their financial discipline and innovation and to give us a signal on what the market price for water should be," says PUB's Chew. According to him, the government maintains demand by signing long-term contracts to buy the water. The long-term planning horizon and the willingness to work collaboratively with universities and private companies mean, says Chew, that "the industry has a chance to flourish, and gives it an opportunity to expand beyond Singapore. This is where the idea of the global hydrohub comes from."

In short, the formidable resources of the Singaporean state have been marshaled to push forward the effort to improve water technology.[30] This is part of a nationwide effort to promote industries related not just to water but also to green buildings and to urbanization. The PUB, which operates the waterworks, won the prestigious Stockholm Industry Water Award in 2007—a testament to the country's strength in water scholarship and innovation. The award was given to the board on the grounds that its approach to water management was "holistic" and "sustainable for different sectors of society in a unique and challenging urban island environment."[31]

One good example of how Singapore both innovates and then rolls out new water technology is the case of NEWater—a technology that uses membranes to make wastewater clean enough to drink. In fact, because the water lacks any minerals after going through the NEWater treatment, it is actually too clean to be suitable for long-term drinking. More than 90 percent of it is used for industrial processes that require clean water—such as semiconductor manufacturing—and the balance is blended with reservoir water before being used as drinking water. NEWater accounts

The reverse osmosis system inside NEWater's plant in Singapore. Photo credit: David McIntyre

for about 30 percent of Singapore's water today, but it is expected to make up 50 percent of the total by 2060. Given the expected doubling in water needs over the next five decades, that will mean a tripling of capacity to reach that target.[32]

There is formidable technology, but even more difficult is convincing the public to back the idea of drinking wastewater. "The 'yuck factor' matters a lot," says PUB chief Chew, no matter how carefully water is treated.[33]

To counteract this, Singapore mounted an extensive public relations campaign that reached as far as religious leaders. An international panel of experts and thousands of tests backed up government assertions that the water would be safe. The campaign climaxed on the country's 2002 national day, celebrating its independence. A crowd of sixty thousand people, led by Prime Minister Goh Chok Tong, drank NEWater from specially produced bottles (the water is not generally available for consumption). "The key thing is that we started with a good product," says Chew. But it was the sort of public relations campaign in which Singapore excels, a top-down campaign that involved a variety of opinion makers, from the prime minister to religious leaders to politicians, business leaders, trade unionists, and the media. "We covered all aspects," says Chew. "We didn't want to take [acceptance] for granted." The country's experience with NEWater demonstrates the importance of a close alliance among technology, implementation, and public relations.

Desalination has been another major focus. One of the world's larger desalination plants, constructed and operated by homegrown water services firm Hyflux, opened in Singapore in September 2013. Research is also under way to reduce the costs of desalination, which historically has been very energy-intensive. When we spoke, Chew was optimistic about a Siemens pilot project that used a different method of desalination (electrochemical) that may use about half the energy of traditional reverse osmosis methods.

Energy use is also a factor in water treatment plants, which account for up to 2 percent of all electricity use worldwide. A project with Suez Environment and a separate one with Keppel Seghers are looking at harvesting the biogas from sewage to produce electricity and minimize the greenhouse gas emitted by sewage.[34] The methane-rich gas produced by sludge already generates about one quarter of the electricity used in Singapore's water treatment plants. The goal is to have it account for 80 percent by 2030.

Singapore has a flexible policy about where its water comes from. Some plants are run by the government and others by private companies.

Olivia Lum is Hyflux's founder as well as its chairman. After growing up in a leaky shack without running water, Lum graduated with an honors degree in chemistry from the National University of Singapore before starting Hydrochem, the precursor to water treatment company Hyflux, in 1989. Photo credit: Hyflux

Two of the four NEWater plants are run by private companies, Sembcorp and Keppel,[35] and two desalination plants are run by Hyflux, which is also private. Keppel and Sembcorp are two of the country's largest conglomerates, both with substantial water-related businesses. But in a country where the state directly or indirectly controls large swathes of the economy, the small, water-only company Hyflux stands out for its entrepreneurialism.[36]

Founder Olivia Lum grew up as an orphan in a Malaysian tin-mining town—ironically, in light of her current business, with no running water.[37] As the tin mines shut and jobs disappeared, workers headed south to Singapore to work in the booming construction industry. Lum, a precocious fifteen-year-old whose schooling had gone as far as it could in small-town Malaysia, joined them, spurred by a school principal who said that she should continue her studies. Turned down by more than ten schools and on the verge of returning to Malaysia, she finally succeeded in gaining admission to a small school.

Lum prospered in Singapore, eventually winning admission to the prestigious National University of Singapore, where she graduated with honors in chemistry. A job at multinational pharmaceutical company Glaxo followed, where her focus was on wastewater research. She was impressed by Glaxo's responsible and pro-active approach to the effluent it produced. "If we needed to treat the waste, we wouldn't hesitate to treat the waste," Lum remembers. "There were not many companies like Glaxo [that would treat wastewater]. No one would want to do that in Asia. You were burning money [if you treated water]. Asia wasn't developed—people were more worried about economic growth than environmental growth." Glaxo's approach prompted her to think more about the region's growing need for clean water.

"I looked around Asia," remembers Lum. "The story repeated itself. Rivers were polluted. The mind-set was that profit was all that mattered. I said 'I am living in Asia. This could be a good business to be in, a sunrise business.' . . . I kind of wanted to save the world."

For most Singaporeans of Lum's generation, who prized the job security and prestige offered by a job at a multinational company like Glaxo, a position like that might have been enough to keep them at the company for decades. But Lum had other ideas. In 1989, she sold her house and car, raising S$20,000 to set up a company called Hydrochem. The firm acted as a sales representative for multinational water treatment companies. Few potential customers in Singapore wanted to buy products from a small business run by a young woman, barely thirty years old at the time, so Lum started traveling to Malaysia on her 110 cc Suzuki motorcycle. She would leave at 5 A.M. to beat the truck traffic across the causeway over the Johor Strait, which separates the two countries, and spend the day peddling her water treatment products throughout southern Malaysia.

Lum did not want to be a trader forever, so she initiated a collaboration with colleagues at her alma mater, the National University of Singapore. Basic water treatment consists of filtration and chemicals, and she says that is just "putting more pollution in the waste. I felt there must be something better." Local water engineers wanted to stick to conventional methods, but Lum focused on membrane technology, a "clean, compact solution" that was on the leading edge of water treatment technology. "People just didn't believe it. Nobody understood," she remembers. Investing heavily in research and development, she reached a milestone when the company began manufacturing its proprietary membranes in 1999.

Two years later, the renamed company Hyflux was listed on the Singapore stock exchange. The company had operated in China since 1994, and the first decade of the 2000s saw expansion in the Middle East and India as Hyflux began building water treatment and desalination plants in addition to selling its membranes. Today, Hyflux has built some of the world's largest desalination plants in locations ranging from Algeria to China to Singapore. Hyflux built Singapore's first NEWater plant, to treat sewage water to a drinking-water standard, and in September 2013, it opened Singapore's largest desalination plant. Hyflux supplies equipment, mostly membrane filters, to plants in four hundred locations. Although most of its business is in Asia, Hyflux sells to projects on every continent except Antarctica. Partners include Mitsui and Hitachi, with whom Hyflux is teaming up to build a $600 million desalination plant in Gujarat, India.

Lum still owns almost one third of the company, and in 2011, she became the first woman and the first Singaporean to win the Ernst & Young World Entrepreneur of the Year award. In 2013, the company reported revenues of $412 million and a net profit of $33 million.[38]

~

The Pearl River Delta has now become the manufacturing workshop of the world. What Manchester and Liverpool were in the nineteenth century, the Pearl River Delta is today, yet on a scale that would have been simply unimaginable in the early days of the Industrial Revolution. The factories there make everything from iPads and iPods to garden gnomes and Barbie dolls to solar panels and electric cars.[39] A ferry that goes up one of the many rivers that feed into the delta ends in Gaoming, home to a major factory of one of the world's most important manufacturers but also one that most people outside of the textile industry have never heard of: Esquel.

Esquel is one of those companies that makes our world of low consumer prices possible. When Esquel set up its Gaoming factory in 1988, there was just a fishing village with some three thousand people. Today, Esquel alone employs forty thousand people in Gaoming and close to sixty thousand in total. Dozens of other companies have joined it in what is now a busy manufacturing city.

Every year, Esquel makes more than one hundred million shirts—in other words, annual production at this single company is almost enough to provide one shirt for every adult American man. It is one of the largest shirt manufacturers in China, which is to say one of the largest in the world.

A Stanford Graduate School of Business case study found that Esquel was the largest Chinese exporter of woven shirts (think button-down dress shirts) and one of the largest exporters of knit shirts (think polo shirts). Among its largest customers are brands like Polo Ralph Lauren, Tommy Hilfiger, Nike, Hugo Boss, and Lacoste. Esquel makes shirts to each company's specifications, putting the company's label on the products. Few mass-market clothing companies in the West own their own factories anymore—it is easier to outsource this business to companies like Esquel.

Headquartered in Hong Kong, the Esquel Group was set up in 1978 by Y. L. Yang, one of the many Shanghainese refugees who fled China after the Communist takeover in 1949 and rebuilt a life in Hong Kong. Esquel' first office was in Tsim Sha Tsui, not far from the terminal for the ferry that takes visitors to Gaoming. The company initially focused on inexpensive clothing and textiles, but as it has grown in size, it has become increasingly sophisticated. More impressive than its $1.3 billion in annual sales is the fact that the company is vertically integrated, from the farm to the retail shop. This is unusual in an industry where companies that make fabric do not usually also cut and sew garments. Chinese laws prohibit Esquel from farming directly, but it buys most of its cotton from long-term suppliers, some of whose drip-irrigation farms it has helped finance. The company has even, on a limited basis, worked with farmers to grow organic cotton. It then gins the cotton (separating seeds from fibers), spins it into yarn, dyes it, and weaves it into fabric that is cut and sewn into shirts. These are then sold through other companies or, to a lesser degree, through its own retail outlets. Most of its operations are in China, from the arid far western province of Xinjiang, where it buys and gins the cotton, to Gaoming and other locations in the Pearl River Delta, where the fabric is dyed, woven, cut, and sewn. It also has factories in Malaysia, Vietnam, Sri Lanka, and Mauritius.

Today, the privately held Esquel Group is chaired by Y. L.'s daughter, Marjorie Yang, who majored in math as an undergraduate at the Massachusetts Institute of Technology (MIT) and went on to get an MBA from Harvard. Margie, as her staff calls her, has served on boards at MIT and at Harvard and Tsinghua Universities. In Hong Kong, she is the chair of the Hong Kong-United States Business Council and the chairman of the Council of Hong Kong Polytechnic University. Esquel's CEO is John Cheh, who holds a PhD in economics from MIT and was a Canadian diplomat and trade official before going into the private sector. Before joining Esquel in 2003, he was president of China operations for plane and train maker Bombardier. Esquel clearly boasts an impressively credentialed management—thanks in large part to its image as a forward-looking company, Esquel is also able to hire enthusiastic and well-educated staff. Among those I met on a visit to Gaoming were employees, all of them Chinese, who had earned advanced degrees from Purdue, Johns Hopkins, and the University of Georgia. "You want to make it fun and make it something people want to do," says Marjorie Yang. "I have been able to attract a great group of people—for example, there was a guy who joined us from a consulting company. He switched companies because he believes in the same kind of ideas we do."

Yang has broad interests in environmental issues. She is a member of the advisory council of the Natural Resources Defense Council (NRDC) in China and chairs the Shan Shui Conservation Center, a nongovernmental organization (NGO) focused on biodiversity and conservation efforts in China's vast interior. From carbon trading to water conservation, Yang sets the tone for the ethos of water and energy frugality that pervades Esquel. "I'm of the 'limits to growth' generation," says Yang. "The Club of Rome report came out when I was at MIT. There is a lot of technology that can be deployed. Use your smarts to do things in a more sustainable way . . . it is not just the environment. It is about sustainable development."

Esquel's sustainability mind-set is built on Yang's conviction that both water and carbon are underpriced. In 2013, energy and water made up about 5 to 6 percent of the company's cost of production, or many tens of millions of dollars. Vast amounts of water are required to grow cotton and then to dye it—Esquel has saved the equivalent of about 160,000 Olympic-sized swimming pools of water in the eight years since conservation efforts started. Yang sees that by increasing the efficiency of its processes, the company has an opportunity to save money immediately and to be in a more competitive position when water and electricity prices rise. So over the past decade the corporate culture has been focused on a continual process of wringing out efficiency gains in water and energy. Making improvements like this is an iterative, never-ending process. It requires a good deal of technical knowledge and enthusiasm all the way down to the spinning mill floor, where the small incremental changes are actually made.

By focusing on efficiency now, Esquel hopes to be ahead of its competitors in a world of higher prices and more limited availability of water and carbon. The initiative to change came from the top but is now part of the corporate culture and appears to be self-perpetuating. "Our thinking is that sustainability means good management and good business," says Agnes Cheng, a company manager who handles sustainability and communications, "and that we are looking into the future, when water and energy may not be heavily subsidized."[40] Typically, there is a short period before the savings on electricity or water pay for the cost of the investments in sustainability, despite the fact that water and electricity for industry are relatively inexpensive in China. This payback period ranges from two months to ten years, says Cheng, though most of the projects have a payback period of fewer than three years.

Given that textiles and apparel are low-margin businesses, savings in energy or water costs have an outsized impact on profits. Recognizing environmental challenges as an opportunity, in 2002 Yang commissioned the Hong Kong think tank Civic Exchange to do a sustainability report for Esquel. Civic Exchange was founded by Christine Loh, a former legislator, one of Hong Kong's most prominent environmentalists, and, since 2012, Hong Kong's undersecretary for the environment. Following the report, in 2005 Yang and Cheh set a goal of reducing water and electricity use by 30 percent—measured per shirt—over the next three years. The target was met. Yang and Cheh kept pushing for further reductions. From 2005 to 2013, energy consumption per shirt fell 34 percent while water consumption fell 52 percent.

Since then, the company has continued to drive down energy and water used per shirt. Today, it uses as much water as it did in 2005, about eight million tons a year, although production has doubled.

Recycling will be increasingly important—the company recycles about twenty-five hundred tons of water a day, or close to one million tons a year. "My basic philosophy is that technology is there to solve problems," says Yang. "When you pollute—anything you waste you are paying for. If you are green and sustainable, you will make more money." Yang would like to see higher water prices and stricter standards, contending that Esquel's efficiency gains are not as valuable as they would be if subsidies were removed. For example, Esquel has developed no-water dyeing techniques. But they are expensive and do not make economic sense because the price of water is still too cheap.

Subsidies, in other words, penalize Esquel by limiting its competitive advantage vis-à-vis competitors who do not invest in water- or energy-efficient technologies. Subsidies also make important innovations like no-water dyeing uneconomical. There may have been a policy case for cheap electricity and cheap water in the earlier years of China's industrial revolution, but now it is slowing the adoption of energy- and water-saving technologies.

In 2009, at the time of the Copenhagen meeting on climate change, Esquel teamed up with Danish-headquartered biotech company Novozymes in an R&D project that reduced water used per unit by 80 percent. "This is an example of how technology and process engineering can change the game," says Cheng, who notes that the accumulation of "hundreds of small things" adds up to significant savings. Other changes include processes that use less bleach when dyeing dark colors and a variety of pretreatment processes that remove wax, cotton lint, and other impurities.

Esquel's sprawling Number 2 Gaoming woven mill makes the massive amount of cloth needed for shirts—a staggering eleven million yards of cloth per year. (If all those five-foot-wide bolts of fabric were placed end to end, they would be long enough to lay a Christo-like carpet from New York City to San Francisco, back across the country to Miami, and all the way up the Atlantic coast to New York City.) Excluding cotton growing, the mill is the most energy-intensive part of the Esquel operations, accounting for two thirds of the water and electricity consumption. It is also, thanks to the fabric dyeing that takes place in the mill, potentially the most polluting. Dyeing is a notoriously dirty process, as dyes often pollute waterways, and the Chinese government estimates that almost 12 percent of China's water pollution comes from textiles.[41] Esquel has worked to reduce the amount of water that it uses in dyeing, intensified recycling efforts, and, where water must be discharged, ensured that it is as clean as possible. Responding to a combination of current laws, a far more aggressive public emboldened by China's blogosphere, and the belief that water use will be increasingly limited, Esquel recycles increasing amounts of water and treats the water that it does discharge to a high standard.

The water recycling and treatment take place near the dyeing mill inside a low-slung compound. On the day I visited, the bubbling wastewater tank was filled with a reddish-purple water that is the effluent from the dyeing process—the color changes depending on the season's fashions. Nearby, dozens of colorful carp swimming in clear water testified to Esquel's recycling efforts, which have been lauded by one of China's leading environmental NGOs in a study that took aim at a number of international brands for their role in buying textiles from polluting Chinese factories.[42] The New York–based NRDC, which has done extensive research into environmental issues in the textile and dyeing industry, also applauds the company's efforts. "Esquel is the gold standard for water and energy efficiency," says Barbara A. Finamore, director of the NRDC's Asia program.[43]

Water is treated up to what Esquel says is a 97 percent clean standard—clean enough that it can be safely and legally discharged. Esquel's discharged water has a Chemical Oxygen Demand (COD) measure of seventy parts per million, a count that is well within the Chinese-mandated standard of one hundred and one that is in line with global standards for similar plants. It is linked to a real-time, always-on system that can be monitored by both Esquel and the government.[44]

Esquel's engineers tell me that their major concern isn't the availability of water at Gaoming, given average annual rainfall of 1,500–2,000 millimeters (almost six feet at the low end). The issue in the future, they say, is likely to be limits on discharges of water. Using less water and recycling more of what is used will be key. Esquel continues to experiment, working now on a reverse osmosis system that will treat up to 70 percent of the water. This should put Esquel in a good competitive position. "We are not only looking at the next three to five years," says Cheng. As a company, "we are more than thirty-five years old and we are looking at the next thirty-five years, or even longer."

A passion at the top of a company matters, and Yang and CEO John Cheh bring that. And although they've been successful in instilling sustainable values in the corporate culture, translating that enthusiasm into sustained action—a process of continuous, incremental innovation—can be difficult after the easy gains of the early years have been made. Once the easy and cheap changes have been made, the question of whether to continue some of the most expensive initiatives—financing drip irrigation for the company's growers, for instance—comes to the fore.

"There has to be a business case," says Marjorie Yang. "We make our living from cotton. Unless you protect underground water, there is no cotton and no cotton textile industry." She notes that in the ten-plus years since Esquel started its sustainability efforts, the company sales and profits have grown dramatically. The Stanford business school case study of Esquel says that Esquel's sales came close to tripling, from $418 million in 2003 to $1.15 billion in 2011. During the same period, the company's net profit grew sixfold, from $21 to $124 million.[44] "My best argument is 'Guess what—we are making money,'" says Yang. "While our results were going up, other people were going out of business."

~

"Whiskey is for drinking and water is for fighting," goes the American saying.

But for Asia, this is no laughing matter. The region needs to use water more wisely, to ensure that good urban and agricultural planning and sound water policies forestall the need to fight for water. For the more arid parts of the region, the choices will be tough. There will be increasing restrictions on water use and a need to use water more efficiently. Governments, notably in China and India, must lead the way by treating water as a strategic commodity on which their nations' continued prosperity, and even survival, depends. Yet they should refrain from the temptation to seize shared water resources unilaterally. Companies and civil society need government leadership in the form of long-term sustainable water policies. Singapore and, in a very different fashion, the Philippines have shown that companies and civil society will respond positively when governments lay down a foundation in the form of good water policies.

7

The Tropical Challenge: *Saving Asia's Lungs*

> We can survive as a business only when the community and the society prosper. In the end it is good business.
>
> —GEORGE TAHIJA

Our van bumps along an unpaved red-dirt road, a road like so many others on tropical Borneo. Although this Texas-sized island has some of the world's largest remaining rain forests, the old-growth trees have been cut where we are driving. It is scrubby country, trees and brush making for a landscape as open to the sun as not. The day is hot, with low clouds and flat light adding to the bleakness of a degraded landscape.

On the left shoulder, a motorcycle is parked, probably an underpowered 175 cc Yamaha or Honda. Two men are just off the road. They have sawed down a tall tropical hardwood tree and are cutting it up with a chainsaw to take it away plank by plank.

Two men, poor—one motorcycle, basic—one chainsaw. The theft will earn them about $40.

This is the mundane reality of illegal logging. This is how forests disappear. People living on the edge of the forest are, quite literally, marginalized. They are poor, usually on the fringes of the cash economy, with little opportunity to earn money. Logging is hard, physically demanding work, and it doesn't pay well. My companions in the van are from the conservation group the Nature Conservancy, and they advise us not to stop to talk to the poachers. Although our group outnumbers the pair, no one wants to confront a man holding a chainsaw.

A felled tree from Indonesia's famed tropical rain forest might be a mere $40 for the motorcycle-riding poachers, but poaching, much of it far more organized, adds up to a business worth somewhere between $30 and $100 billion each year. The estimates for illegal logging vary wildly, but the illegal trade accounts for somewhere between 10 and 30 percent of the total global trade. In some countries, the percentage of trade that's illegal is much higher—as much as 90 percent. In Indonesia, which includes most of Borneo, it accounts for an estimated 40 to 80 percent of the total trade.[1]

Timber poaching is a serious threat to the environment. Deforestation, burning, and land clearing, mostly of tropical rain forests like the one in Borneo, contribute 17 percent of all man-made greenhouse gas emissions—more than transport (13 percent) and almost as much as industry (19 percent).[2] There are three places in the world where activity is concentrated: the Amazon basin (mostly Brazil), Central Africa (especially the Congo), and Southeast Asia (especially Indonesia). It would be naïve and overly optimistic to say that the battle against deforestation is going well—indeed, in Indonesia deforestation has accelerated since 2000.[3] But there is a way forward that will protect forests even as economic opportunities for farmers and forest dwellers are created.

~

Forests are the world's lungs. Our lungs use oxygen and produce carbon dioxide as a waste product; in contrast, plants use photosynthesis to bind carbon dioxide into an organic compound and remove it from the atmosphere. As humans burn more carbon-based products, such as coal, it is even more important that forests be available. Tropical forests are especially important because they grow faster than those in the boreal regions of Europe, Canada, and Russia and so are more efficient carbon sinks—better able to sponge up carbon dioxide and slow the buildup of greenhouse gases. As forests disappear, these sinks no longer have enough capacity to carry all this carbon away.

The role that forests play in our global climate is often underestimated, and cutting down trees seems more like an aesthetic affront than an act of climate violence. But clear-cutting tropical forest is a kind of global terror in the climate wars. The influential *Stern Review*, released in 2006, noted that reducing deforestation is the "single largest opportunity for cost-effective and immediate reduction of carbon emissions."[4]

Logging and monoculture plantations are problematic not just because they destroy living trees and reduce the size of the carbon sink; somewhat counterintuitively, the related destruction of what is already dead is also damaging. Many of Indonesia's forests are on peat soil, the decayed remains of centuries of forest. Peat is an important carbon sink—it stores carbon that would otherwise be released into the atmosphere. When it is disturbed, peat releases much of that stored carbon, meaning that badly practiced tropical logging contributes to a disproportionate amount of carbon emissions. An estimated 70 percent of tropical peat is in Southeast Asia. This problem is not, however, limited to the tropics. Indeed, Britain is moving to ban peat in backyard gardening by 2020, largely because of concerns about peat's role in contributing to greenhouse gas emissions.[5]

Indonesia is already one of the world's largest carbon emitters despite being relatively underindustrialized and largely agricultural. Part of Indonesia's global carbon footprint reflects its vast size and large population—it has about 250 million people, trailing only China, India, and the United States. Indonesia is a relatively poor country, despite its population placing a distant sixteenth in global economic rankings. Unfortunately, it ranks much higher on the list of carbon emitters, with one estimate putting it as the world's third-largest carbon emitter and another pegging it as the fifth-largest. It's these deforestation activities that make its carbon footprint so large.

Forests are not only important in moderating the pace of climate change; they are also the home to many of the Earth's species and are key to our planet's legacy of biodiversity. Tropical rain forests are especially rich sources of biodiversity, but this is under threat from palm oil plantations and logging. Already, most Asian countries have lost 70 to 90 percent of their wildlife habitats. In Indonesia, more than one million hectares of forest (an area almost as large as Connecticut) were destroyed by fires in 1997 alone. The choking, smoky haze that accompanied the fires affected seventy million people in six countries. Similar losses occurred in July 2013. Many of those fires were set deliberately as a way of clearing land, much of it already degraded, for palm oil plantations. Deforestation also means more flooding and devastation in the wake of storms and makes providing drinking water that much more difficult.

The importance of Indonesia's forests and the danger they are in can hardly be overstated. They are the largest and most biologically diverse in the world after those in Brazil and the Congo, but they could be effectively gone in a decade, warns a report by the United Nations Environment

Program (UNEP). More than one third of Indonesia's forest cover was destroyed during the twentieth century, reducing it from an estimated 170 million hectares to 98 million. Half of what remained in 2000 was estimated to have been degraded.

The rate of deforestation is increasing. The voracious demand for Indonesia's trees comes from several industries. Indonesia's pulp and paper industry is one such source. It has seen capacity increase more than tenfold since 1990, and its paper and pulp mills require new swathes of forest to provide raw materials. Industries that rely on wood are another source of demand. Indonesian trees end up as furniture in the United States, chopsticks in Japan, and construction materials in China.

Agriculture, especially palm oil, is the single most important factor. An unholy partnership between loggers and farmers throughout much of the tropical world facilitates this conversion from jungle to plantation. Loggers, often with the complicity of local officials, cut tropical hardwood trees, the soul of the jungle. The degraded land is cleared, often by burning—itself a source of environmental damage—and turned into monoculture plantations.

Converting forests into monoculture palm plantations is, from a short-term financial standpoint, doubly beneficial. The timber is sold, and then the land is ready for a plantation. From an environmental perspective, however, palm plantations are doubly damaging. The need for plantation land contributes to clear-cutting forests. Then, once the plantations are in operation, they are often managed unsustainably, using poor agricultural practices that lead to loss of biodiversity. Other environmental damage comes from the burning of waste material, effluent runoff, and the heavy use of fertilizers and pesticides.

The demand for palm oil is likely to continue increasing. Palm is a miracle plant, with its caloric output per hectare being a multiple of the outputs of other food oils. Palm oil is the world's most popular edible oil, accounting for one third of total edible oil, comfortably ahead of soybean oil (27 percent). Palm oil and its more valuable sibling, palm kernel oil, are produced from trees grown on plantations that still use methods pioneered by colonial British planters in nineteenth-century Malaya. The trees that produce these oils are extraordinarily productive, yielding more than eight times as much oil per hectare as soybeans. This high yield means that less land is needed to meet demand; palm oil accounts for only 5 percent of the land used to grow edible oil globally, far behind that used for soybeans, rapeseed, cottonseed, and sunflower seed.

Although palm oil is widely used throughout the world, China and India and their increasingly wealthy populations account for about two thirds of its global consumption, and demand is growing. The two biggest uses, in household cooking and the food industry, each saw its palm oil consumption double in the first decade of the century. Palm oil and palm kernel oil are used in everything from candy to shampoo, from biofuel to cooking oil. Although Western consumers are unlikely to come across palm oil in a stand-alone form, it is in many foods and consumer products.

Palm is also used in biofuels. The demand for palm and other food staples used for biofuel, artificially aided by European Union and United States' subsidies, also acts to push up food prices, although there is little agreement as to how much. Palm oil used for biofuels jumped from zero to 1.8 million metric tons in just six years, from 2003–2004 to 2009–2010.[6]

Indonesia and neighboring Malaysia are ideally suited for palm cultivation. About 85 percent of the world's palm oil comes from these two countries, with Indonesia's fast-growing palm plantations alone supplying more than half of the world's production. In Kalimantan, Indonesia's part of Borneo, oil palm plantations grew from 903 square kilometers in 1990 to 31,640 square kilometers in 2010—in other words, from an area a bit bigger than New York City to one about the size of Maryland. There are some consumer groups, however, that are trying to make a dent in this continued growth.

A bored office worker reaches for a Nestlé Kit-Kat in a parody commercial. As our mind-numbed antihero tears open the candy wrapper, his colleagues see what the office worker does not—that the chocolate is coating not a Kit-Kat but an orangutan's finger. Sounds of the office give way to those of the jungle, but birdcalls are drowned out by the whine of chainsaws and a tree crashing to the forest floor. The oblivious clerk bites into the candy, blood dribbling down his chin and onto his computer keyboard while his horrified coworkers look on. A plea to "Give the orangutan a break" is followed by a silent shot of land that has been cleared for a palm oil plantation and this pitch: "Stop Nestlé buying palm oil from companies that destroy the rainforests." The advertisement ends with a link to a website where viewers can take further action.

Nestlé claimed that the advertisement, released by Greenpeace in 2010, was unfair, but it was brutally effective.[7] As the success of the campaign

Greenpeace ran a parody KitKat ad that depicted a bored office worker taking a bite of the popular candy—which, to the horror of his coworkers, was in fact the finger of an orangutan, a species threatened by forest clearing. The ad served to highlight the use of unsustainable forest clearing in producing palm oil, an ingredient used in Nestlé's products. The company now has a goal of, by 2015, using only palm oil certified as sustainable. Photo credit: © Greenpeace International

illustrates, consumer pressure is probably the best hope to improve forestry practices. The food industry—where much of the palm oil goes—is unusually vulnerable to consumer boycotts. In a world of commodity food and widespread choice, even a sniff of consumer doubt about a product can cause sales to plummet. As palm oil production increased, so, too, did the backlash from environmentalists and consumers, angry that demand for palm oil was fueling the destruction of tropical rain forests. The industry's answer has been to set up a voluntary body designed to enforce at least minimal standards.

The Roundtable on Sustainable Palm Oil was inaugurated in 2004 and includes everyone from palm oil growers to commodities wholesalers, such as Cargill and Wilmar, to multinational food companies, such as Nestlé and Unilever, to retailers, such as Wal-Mart and Carrefour. The roundtable mandates a set of prescribed practices among those growers it certifies. These range from prohibitions on planting on recently cleared forestland and peat bogs to requirements for sustainable farming practices, such as water and pest management and minimization of soil erosion. By the end of 2012, 15 percent of palm oil (about 8.2 million metric tons) was certified as compliant with the roundtable principles.

When the Kit-Kat spoof was released, Nestlé's first reaction was to attempt to have the video removed from YouTube. That only attracted more attention and provoked a wave of criticism. Two months later, in May 2010, Nestlé announced that it would no longer buy palm oil from companies that own or manage "high-risk plantations or farms linked to deforestation." José Lopez, a Nestlé executive vice president, told a gathering in Kuala Lumpur that although Nestlé bought only 0.7 percent of global palm oil, it was committed to a variety of measures to stop tropical deforestation. Nestlé said further that it was calling for a moratorium on rain forest destruction for palm oil in Indonesia; that it had joined the Roundtable on Sustainable Palm Oil; that it had suspended purchases from an unnamed supplier (widely believed to be Golden Agri-Resources, an affiliate of Indonesia's Sinar Mas, another target of Greenpeace videos); that it was working with the nonprofit Forest Trust; and that it would buy only sustainably produced palm oil by 2015.[8] Later Nestlé also pledged to protect all high-carbon-value sites including, explicitly, peat lands and high carbon-stock forests as well as the rights of indigenous local people. In 2013, Nestlé achieved its goal of buying only palm oil that had been certified as sustainable by the roundtable, two years ahead of its original commitment.[9] In concert with these changes, the company is switching to more sustainably grown coffee beans for its Nescafe and high-end Nespresso products and is making its Nescafe factories more energy-efficient.

Other large Western food companies have found the Roundtable on Sustainable Palm Oil and other initiatives to be useful in setting sustainability goals. In 2010, Unilever, partly in response to orangutan-dressed protestors at its offices in London, Rome, and Rotterdam, pledged to use only sustainable-certified oil by 2015 and to provide full traceability of its palm oil by 2020. Cargill announced that all of its Indonesian operations have been certified by the roundtable and that it was working with smaller suppliers to help them achieve certification.[10]

In an interesting reflection of where the demand for certified oil is coming from—namely, Western consumers—Cargill promised that all palm oil sold to Europe, the United States, Australia, and New Zealand will be certified and/or bought from small landowners by 2015; the pledge will be worldwide by 2020.[11] Western countries have started to follow suit; six European countries (the United Kingdom, Netherlands, France, Belgium, Switzerland, and Germany) have committed to using 100 percent roundtable-certified palm oil in the near future; the roundtable lists the United States, China, India, Australia, and Italy as "nations showing positive momentum."

As Cargill's geographic segmentation demonstrates, it is largely consumers in the West that are pushing for more sustainable practices; there has been little demand for sustainability to date from Asian consumers. That may start to shift. India is the world's largest market for palm oil, accounting for 23 percent of global consumption in 2011–2012, and in 2013, World Wildlife Fund (WWF) India launched a campaign to encourage sustainable palm oil consumption. This is a small start—but a start nonetheless. Without pressure from consumers in India and China, progress will be illusory: certified plantations would be able to sell to the West, and those that contribute to deforestation, forest degradation, and unsustainable practices would sell to the Chinese and Indians in a destructive race to the bottom.

That change is being driven by Western multinationals, prodded by their customers, parallels the experience within other industries, such as clothing and computers. Nike spent much of the 1990s saying that conditions at its supplier factories were basically fine and, besides, it did not control these factories, so it had minimal leverage. After college students and other important consumers attacked Nike and the company was ridiculed by the *Doonesbury* cartoon strip, the company forced changes on its suppliers that improved working and environmental conditions. Apple's situation was similar. Like Nike, Apple does not own its production facilities. But when a spate of worker suicides occurred in 2011 at Foxconn, a Taiwan-based manufacturer with massive operations in mainland China and Apple's largest manufacturer, Foxconn and Apple were forced to react with higher wages and improved worker and environmental protection.

There are real limitations to this voluntary approach. Critics have pointed to the Roundtable on Sustainable Palm Oil's lack of enforcement mechanisms. Indeed, it seems that the most influential sanction will be the power of the purse—a decision to stop buying from a supplier. But some combination of the two may start to make an impact. Indonesia's Sinar Mas Group, in the past widely considered by environmentalists to be one of the most environmentally destructive major business groups in the country, has slowly started to sign on to industry codes, notably with an early 2013 agreement to improve its timber practices. The giant Wilmar International Group, a Singapore-based $45 billion commodities wholesaler that is one of the most important players in the palm business, responded to pressure by promising at the end of 2013 to source palm oil sustainably. Wilmar had found itself under attack on issues from orangutan protection to rights of indigenous people and earlier had joined the roundtable. Being inside such a group, coupled with its public commitment to improve practices, provides a strong impetus for the company to improve its standards. The Consumer Goods Forum, a broad international grouping of some four hundred companies, is also important and has pledged to achieve zero net deforestation by 2020. However, the pace of change is frustratingly slow, and there are very clear limits to how much can be accomplished by voluntary industry efforts alone.[12]

~

The sun always rises just before 6:00 A.M. on Belitung. And it is shortly after sunrise, six days a week, that managers, foremen, and harvesters at ANJ Agri's Ladong Jaya palm plantation line up for morning musters, following routines that go back to British planters in nineteenth-century Malaya. Indeed, the meticulously maintained soccer field where harvesters group around the foremen helps give the place something of a British air, a feeling reinforced by the matching ANJ Agri short-sleeved polo shirts, high socks, and shorts that the managers and assistant managers wear. Simple, one-story houses with red-tile roofs ring the far side of the soccer pitch. A school bus filled to capacity with white-shirted primary school students passes by. Of course, there are other clear signs that this is not England. Instead of a church, there is a mosque on the far side of the field. And the trees surrounding the settlement are palms and tropical hardwoods.

It is in places like Ladong Jaya that the real test of improved practices comes. To get a better sense of the realities of sustainable tropical monoculture—or to find out if the very idea of a sustainable palm plantation was perhaps an oxymoron—I had flown an hour north of Jakarta to Belitung, an island 3 degrees south of the equator off the coast of Sumatra. Belitung was once known as Billiton, and it was here that one of the companies that today is part of mining giant BHP Billiton got its start with the 1851 discovery of major tin deposits. Although the degraded open pits of abandoned tin mines still dot the landscape, this island, once famed for tin, is now dominated by palm. ANJ Agri, the owner of the planation I am visiting, is a Jakarta-based company that has about fourteen thousand hectares of palm under cultivation on Belitung. With an area that is about the size of New York City's Staten Island, the plantation is sizable, although not large by Indonesian standards.

George Tahija, who founded palm producer Austindo Nusantara Jaya, in a new plantation development area in south Sumatra, Indonesia. Photo credit: George Tahija

ANJ Agri is a publicly traded company (its 2013 initial public offering on the Jakarta stock exchange was managed by Morgan Stanley); however, the founding Tahija family's continuing personal involvement gives the company an unusual character. The family is well respected in Indonesia. Patriarch Julius was a World War II hero and a prominent figure during the independence movement that followed.[13] One of Julius's two sons, George, is involved with the family businesses on a day-to-day basis. He is also a mountain climber, a sailor, an environmentalist, and a nature photographer whose photo books document many of Indonesia's wild places. Tahija's nonprofit activities have included significant involvement with the WWF and the Nature Conservancy. George and his brother, Sjakon, an ophthalmologist, are also involved in a number of health-related philanthropic projects.

George Tahija has a vision for the company as "a world class sustainable food and energy company which elevates the status of the Indonesian people." Note that the vision is not for environment at the expense of development but for environmental protection that contributes to development.

There are about 120 harvesters at ANJ Agri's Ladong Jaya plantation (and thousands throughout the country). Local residents generally don't want to endure six-day workweeks for average pay of about $200 a month, so most of these workers are from Java, Indonesia's most populous island, and have nothing but an elementary school education. Although life is improving for plantation workers, reflected in the growing number of motorcycles at their company-provided housing, these men do not have the luxury to spend a lot of time thinking about tropical rain forest destruction. Yet they are the ones tasked with helping to stop the most destructive monoculture practices.

Leading the musters at the ANJ Agri plantation when I visited was Joseph Gomez, a senior plantation manager just shy of his sixtieth birthday. Gomez, an ethnic Indian from Malaysia, has been in the plantation business for close to forty years, most of it at the venerable British-turned-Malaysian company Guthrie, a leader in palm plantations. When Gomez started working in the early 1970s, there was no concern with sustainability. As he describes it, old-growth forests were cut and then burned to clear the land for new plantations. Like most commercial agriculture operations, palm plantations generate more waste than usable product; after harvesting the palms, the waste was simply burned, and the milling effluent, the so-called black liquor, was dumped into large holding ponds. Philip Liu, in charge of several of ANJ Agri's plantations, is another Malaysian who also started at Guthrie. As he recalls, "Even then we knew it was bad. After you came back from supervising burning, you had trouble breathing."

Today, in line with ANJ Agri's Roundtable on Sustainable Palm Oil certification, burning is forbidden. Instead, waste palm fronds are used as mulch, helping to prevent soil erosion. The goal is zero waste, with an emphasis on reuse, recycling, and natural methods. Besides the palm fronds, the waste includes empty fruit bunches, the pulpy fibrous leftovers after the oil has been pressed. Like the fronds, these now are put back on the ground, serving as natural compost.

What to do with the black liquor, the waste effluent from the palm oil milling process, is a bigger challenge. At ANJ Agri, the answer comes in the form of a significant investment to convert waste into energy.

A new $17.5 million biogas plant was being tested when I visited. From the outside, the plant looks something like covered tennis courts; inside, three huge fermenting tanks mix the black liquor waste from the mill with specialized yeast. The resulting anaerobic fermentation produces methane, which in turn is burned to power electrical generators. The plant should pay back its cost in about three years by allowing ANJ to save on the diesel fuel now used to power the mill as well as by providing it with excess power, which it can then sell back to the state-owned utility. The remaining sludge will be used as fertilizer. With fertilizer accounting for up to 60 percent of operating costs (labor is only about 10 percent), any saving on fertilizer is significant. A smaller biomass plant burns the kernels' shells to produce more generator power.

ANJ Agri also has come up with a natural way to combat pests. The company encourages the breeding of barn owls and now has over a thousand among its 14,229 hectares of palm on Belitung (about one -tenth of ANJ's total plantation area in Indonesia). The owls hunt rats, which can cause crop losses of up to 10 percent, and eliminate the need for environmentally destructive pesticides. The need for herbicides is cut by planting crops that attract caterpillar-eating insects.

These are not highly sophisticated processes but rather are good farming practices on a large scale. Even so, they still take time and trouble and commitment from the top. Changing established practices requires a combination of knowledge, money, and time—resources that the many small holders selling palm oil don't have. Ideally, a national agricultural extension service would help spread best practices, especially to the small farmers whose production makes up one third or more of the total bought by larger food processors. However, the reality on the ground in Indonesia is one of such limited governance and government capacity that an extension service is little more than an idealistic dream. In a nation of 250 million people spread across a sprawling archipelago of seventeen thousand islands with per capita annual incomes of only about $3500, even the best government would struggle to impose fair, efficient rule, let alone best-practice farming techniques. A consensus-driven industry body with no power to sanction offenders, like the Roundtable on Sustainable Palm Oil, has many weaknesses. But it is hard to see a more effective alternative for now in Indonesia.

Shortly after I visited Belitung in early 2013, ANJ Agri raised about $45 million for the company's expansion in an initial public offering that valued the company at about $450 million. After the IPO, I asked George Tahija what part of the sustainability effort he was proudest of. Whereas many executives might have used this as an invitation to indulge in self-congratulation, Tahija's answer was revealing, focusing as it did on continuing difficulties.

"Frankly, I find it one continuous struggle," Tahija said of his efforts to ingrain sustainable cultivation practices. "To be frank, I am not best practice yet. My biggest challenge is that every effort needs a champion—if I remove myself from the scene, old habits may revert. It is the efforts we have done to build the right corporate culture. All the things you saw in the field are the result of that corporate culture. I wouldn't dare to say that our corporate culture is sustainable. We definitely have good corporate culture. I am most proud of our corporate culture but at the same time I recognize the fragility of it."

For now, ANJ Agri's sustainability approach owes much to the enlightened paternalism of the Tahija family. Repeatedly, when I asked about the reason for something, staff told me that it was at the request of "Pak George," using an Indonesian term of respect for Tahija. Should palm trees be planted even at the cost of disturbing some seventeenth-century royal graves? Pak George decided to preserve the graves, and today they remain a forest sanctuary where visitors are serenaded by the extraordinary orchestral music of rain forest insects. A new training center for workers was under construction when I visited. Worker housing, provided free of charge, appeared clean and well maintained. Where did the inspiration come from? Pak George, of course. "It's been part of the family philosophy that wherever we do business, we should be seen as an asset to society and the community and that we should not be a burden," Tahija told me later. "We can survive as a business only when the community and the society prosper. In the end it is good business."

ANJ's Tahija family exemplifies an important trend in Asia's sustainability movement, whereby long-term family owners have the vision—along with the capital, technology, and managerial resources of any large company—to be pioneers in sustainability efforts. Environmental sustainability initiatives at these companies invariably are part of a broader engagement with the communities and countries in which they operate. These initiatives reflect a variety of different motives, from a paternalistic sense of responsibility to a sense of doing what is right to a more pragmatic need to maintain good relations with local people, politicians, and others in order that the company can enjoy its social license to operate. It will be important to see if the examples of these leaders inspire or encourage other business leaders to act in a similar fashion.

~

Great Giant Pineapple is everything its name promises: the company says its thirty-thousand-hectare farm on the Indonesian island of Sumatra is the world's largest pineapple plantation—it is five times the size of Manhattan. Although it started planting its pineapples only in 1979, the company calculates its total pineapple output trails only Dole and Del Monte, with its annual production about two thirds that of market leader Dole.

Already, Great Giant claims to run the world's largest pineapple factory. More than one million pineapples are peeled, sliced, and canned every day of the year except during the Muslim holiday of Ramadan. Those pineapples are shipped around the world, bearing the labels of more than six hundred different customers. If it is not a Dole or Del Monte canned pineapple, there is a pretty good chance it is one of Great Giant's. Each year, the company produces 500,000 metric tons of canned pineapple and pineapple juice—about one can of pineapples for every American. By 2020, it aims to be the world's largest pineapple company.[14]

Great Giant is a good example of a company that has adopted environmentally sound farming practices for practical business reasons. Pineapple waste left after canning is mixed with feed for the company's twenty-five thousand cattle; cattle manure is used for fertilizer, in turn cutting the need for chemical fertilizer. After harvesting, the pineapple plants are mulched and used for compost, rather than being burned, as they were in the past. Bamboo trees planted along ponds and wetlands protect against soil erosion; the bamboo is sold as chopsticks and as paper and is used as mulch for the pineapple fields. "Solid waste is a closed loop," says Ruslan Krisno, who runs the company's sustainability unit, "from mulch to pineapple plant to fruit to cows to mulch."

Great Giant Pineapple illustrates how a sustainability commitment from senior management prompted a numbers-based approach to a broad range of sustainability issues. This in turn led to better business practices. What started out as a fairly narrow approach regarding sustainability forced managers and lower-level employees to rethink their business-as-usual approach. What emerged is improved efficiency as well as a more environmentally benign firm, one that has annual sales of $350 million and fifteen thousand employees.

These far-reaching sustainability efforts began in the 1980s when Great Giant struggled to solve the problem of the goopy pineapple canning waste. "We had the waste, and we didn't know what to do with it," remembers Krisno. "If you throw it away, you will get complaints from the community." So the company bought cattle to avoid dumping the waste. "This project was started to solve environmental issues," says Krisno. "It turned out to be a business." The sustainability unit was established in 2011 at the initiative

Husodo Angkosubroto, the head of Great Giant Pineapple, in one of the company's fields in Sumatra, Indonesia. Photo credit: Great Giant Pineapple

of the group's chairman, Husodo Angkosubroto. The hope was that this push would help the company run more efficiently and build loyalties among its buyers—as well as allowing the company to command a price premium for the recognition of its sustainable practices (a reminder of the importance of consumers in catalyzing change in the food business). That the company is trying to use resources more wisely reflects a concern over scarcer resources as well as social pressure from local farmers, politicians, and Western consumers.

After a high-level commitment from management, one of the most important prerequisites for ingraining more sustainable business practices is measurement. Great Giant Pineapple has embraced a variety of different measurement tools. After setting up the sustainability unit, it commissioned a detailed energy audit as well as a separate measurement of its carbon footprint. In 2012, it started the process of getting ISO 15001 status, a recognition of good energy management. In the future, it intends to be audited under the Global Reporting Initiative, a rigorous environmental and social sustainability framework. This sort of intensive auditing also gives the company more traceability and more certainty over its food chain. Because it owns all of its production, it will be able to tell buyers where a can of pineapple came from, down to the individual plot, and when it was canned. In an era of increasing concern about food safety, this makes good business sense. The company also has a goal of achieving carbon-neutral status by 2015. "We have to be doing this before the consumer or our customers are demanding it," says Krisno.

When Great Giant Pineapple set up its sustainability unit in 2011, renewable energy was absent from the fuel mix. That year it built a $4 million biogas facility with the help of Global Water Engineering, a Belgian company. The biogas facility takes that liquid waste from pineapple canning and produces about 1.8–2.0 MW of power, reducing the need for burning coal at a captive power plant. By early 2014, this biogas facility produced 7 percent of the company's energy, and palm kernels provided another 10 percent, moving the company more than halfway toward a management target of generating 30 percent of its electricity from renewable sources by 2016. The biogas has allowed the company to cut the use of heavy fuel oil

Great Giant Pineapple picks and processes over one million pineapples a day at its southern Sumatra pineapple plantation, the world's largest. Photo credit: Mark L. Clifford

and coal and to sell carbon credits under the Kyoto Protocol. The roughly $200,000 a year that it receives for selling the carbon credits, which allow foreign buyers to offset their carbon emissions by helping finance Great Giant's cuts, is not significant, but at least it allows the pineapple company to pay for the cost of verification and inspection, ensuring that it is accurately measuring what it is doing.

Besides saving on coal, the biogas operation has allowed Great Giant to cut the land required for its effluent holding ponds from twenty hectares to one hectare. The company has also eschewed well water in favor of rainwater collected in holding ponds. This is environmentally more benign but also saves on the electricity and equipment costs associated with drilling for water. The company also plans to cut the use of chemical fertilizer, which accounts for about one third of the operating costs of the pineapple plantation (about the same percentage as labor costs), by 40 percent from 2014 to 2020, thanks in large part to a $14 million organic fertilizer plant. In short, the decision from the top to embrace sustainability set in motion an internal process that resulted in incremental innovation, long-term cost savings, and a stronger competitive position in global markets.[15]

By the company's own reckoning, there is a ways to go. Krisno says that the pineapple industry is not nearly as sophisticated as the palm industry, with its Roundtable on Sustainable Palm Oil standard. Great Giant Pineapple doesn't yet know the total amount of water and carbon used to produce each can of pineapples, for example. Early though it is in this process, the commitment to manage its resources more efficiently is being woven into the fabric of its corporate culture.

~

The entire world has an interest in the preservation of tropical forests. If Brazil, the Congo, and Indonesia exercise their sovereign resource rights and cut down rain forests, they will enjoy short-term economic benefits. What right do outsiders have to tell these countries not to develop? Yet destruction of the tropical rain forests would mean the entire world would be losing an important carbon sink and a rich source of biodiversity. Given that the rich world is responsible for much of the carbon already present in the atmosphere, as a result of two centuries of carbon-led development, could wealthier nations pay poorer ones to be environmental stewards?

The belief that the rich world can (and, many people would say, should) bear much of the financial responsibility for preserving the environment underpins the thinking behind what is called REDD+, an ambitious and far-reaching forest protection plan. The story of REDD started far from the tropical forests of Borneo with a 2008 United Nations plan aimed at giving forest-rich developing nations a financial incentive to preserve their forests and take more sustainable, low-carbon development paths. The program received a big boost in 2009 when Prince Charles hosted a private meeting at St. James Palace. Some of the world's most powerful people were among the attendees, including U.S. Secretary of State Hillary Clinton, German Chancellor Angela Merkel, World Bank President Robert Zoellick, European Union President José Manuel Barroso, and Japanese Prime Minister Taro Aso. The group agreed that urgent action was needed to halt the alarming rates of deforestation. This meeting was the genesis for an initiative formally launched in 2010 in Oslo with the ungainly name of Reducing Emissions from Deforestation and Forest Degradation, or REDD. The idea is to pay countries to be responsible forestry stewards. An enhanced version, which includes a greater role for conservation and sustainable management of forests, has been named REDD+.

REDD+ rests on a simple concept, one that reflects the belief that markets can work better than heavy-handed government initiatives. Simply put, the forests are worth more alive than dead—clear-cutting and ill-conceived monoculture are less profitable in the long run for the world as a whole than preserving the carbon-absorbing capacity and biodiversity of tropical forests. To meet the very real income needs of people in developing nations, the rich world will pay countries like Indonesia to keep their forests untouched or to harvest them in a sustainable fashion.

REDD+ got off to a promising start. By March 2012, $8 billion had been pledged by rich countries and multilateral institutions like the World Bank. The promise was that by 2020 an impressive $100 billion would be devoted to REDD+ efforts.

Indonesia has been at the forefront of REDD+. In October 2009, Indonesia announced an ambitious plan to reduce greenhouse gas emissions by 26 percent, compared to a baseline, business-as-usual scenario. Indeed, President Susilo Bambang Yudhoyono promised that the cuts could be as high as 41 percent if international financing assistance was available (such a reduction would be equivalent to 8 percent of the total global reduction recommended by the United Nations Intergovernmental Panel on Climate Change).[16] Given that almost 80 percent of Indonesia's greenhouse gas emissions are from deforestation, the program was designed to have a globally significant impact.

The international community certainly took notice. On May 26, 2010, Norway promised to pay up to $1 billion to Indonesia to support what the Norwegian government said was "the largest absolute reduction commitment made by any developing country."[17] Indonesia, advised by consultants at McKinsey & Co., promised to put a two-year moratorium on new logging concessions for its part of the sweeping REDD+ program. The program is intended to initially fund scientific mitigation studies, followed by a pilot greenhouse gas reduction/mitigation program. The idea is that by 2018, eight years after the signing of the agreement, such a program will be rolled out nationally. Illegal logging, which the Norwegian government says costs Indonesia $2 billion a year in lost revenue, taxes, and duties, will be reduced as a result of better management of the country's forests.

However, four years into the program, when President Yudhoyono ended his second and final term, REDD+ results have been disappointing. Although the moratorium on new logging was extended in May 2014 for two more years, this theoretically elegant program has only a slim chance of success in a country where vague property rights, weak rule of law, and widespread corruption are coupled with continued bickering over leadership of the REDD+ program.[18] "It is a good idea," says George Tahija of ANJ Agri, which had hoped to take part in the program but gave up because of impediments at the local level. "The implementation, the successful implementation, depends on the clarity of the implementing regulations," he adds. "And I haven't seen them."

~

How do good ideas hatched in the elegant rooms of St. James Palace—or even at Greenpeace headquarters—stop a couple of poor guys with a chainsaw and a motorcycle in Borneo from cutting down a tropical hardwood tree to make a much-needed $40?

Indonesia's vibrant Jakarta-centric civil society and media may someday play a more important role in battling environmental misdeeds in that large, geographically splintered country. Until Suharto's three-decade-long autocratic rule ended in 1998, the media and civil society were muzzled. Now, in an era of social media that empower activists using tools as diverse as satellite photographs and on-the-ground video, a torrent of information is available. This information could go a long way to counteract the power of local political and business interests.

For now, voluntary efforts like the Roundtable on Sustainable Palm Oil, buttressed with global consumer pressure designed to force agricultural producers to improve their environmental practices, are likely to be more powerful forces for change than multilateral or government-led initiatives. Asia Pulp and Paper and its sister palm oil company, Golden Agri-Resources, have willingly submitted to a variety of certification efforts, and palm giant Wilmar has entered into a similar agreement; these are powerful testaments to the growing ability of an international network of activists to partner successfully with those in the industry. Strong consumer pressure on multinational companies has forced these companies to cut off purchases from environmental outlaws. The need to protect valuable international brands in turn is forcing change all the way to the forest floor in places where media, NGOs, and the government are weak. This successful transnational consumer activism is one of the more hopeful developments in what remains an appalling state of affairs.

8

"Adhere and Prosper"
One Company's Quest for Green Power

If we do nothing but coal we will be crucified.

—ANDREW BRANDLER

What does a company do when it faces the prospect that its traditional business will disappear? Former photo film maker Kodak and tech giant IBM are just two of the many companies that have faced radical changes in their traditional businesses, changes that they have met with varying degrees of success. The challenge of a seismic shift is much bigger in a capital-intensive business like electrical power generation. Each power plant typically represents an investment of several billion dollars and is meant to stay in service for many decades, often a half-century or more. The prospect that these assets could be stranded, and thus made worthless, by changing environmental legislation or better electrical-generation technologies is a powerful incentive for owners to try to plan decades into the future.

With respect to environmental matters, few utilities have started to plan for a low-carbon future more thoughtfully and effectively than Hong Kong–based CLP Holdings. CLP's 1990s ambition to acquire power plants throughout East Asia, South Asia, and Australia led to a dilemma: most assets for sale were coal-fired power plants, but the world needed to move to a lower-carbon future over the next few decades.

Asia's growing electricity demand, most of it still met by coal-fired power plants, is the single most important factor in determining both the continent's carbon emissions and its air pollution levels. No utility in the region, and arguably none in the world, has dealt with this dilemma as forthrightly as CLP, which in 2007 pledged to reduce the carbon intensity of its electricity production by more than 75 percent by 2050. This chapter deals at length with CLP's story, a story that holds broader lessons for business leaders and government policy makers as they move to cut carbon emissions and reduce pollution.

~

Michael Kadoorie is one of Asia's richest men, with a family fortune of over $5 billion. He is scion of the Kadoorie family, Iraqi Jews whose grandfather came from Baghdad to Hong Kong in 1880 and who grew rich supplying electricity to Hong Kong and creating the luxury Peninsula hotel chain. The family is among the most successful of the outsiders who came to Asia in the nineteenth century to make their fortune, and its electricity company is one of Asia's largest. Yet Kadoorie now aims to turn his family's coal-burning utility into one of the world's greener power companies.

Kadoorie's 2007 pledge that CLP Holdings will largely stop releasing greenhouse gases by midcentury makes the company unique among the fraternity of major power companies. It started out as a utility operating in a forgotten corner of Hong Kong and has ended up with about 80 percent of the city's electricity customers, serving some 5.7 million. It has sales of more than $13 billion and owns assets of more than $27 billion. It is the larger of Hong Kong's two electricity companies, providing power to more than two thirds of the city. The company is one of Asia's largest privately owned utilities. It is also one of the best run. Energy expert Platts awarded it the number one ranking for financial performance among Asian utilities in 2009.[1]

CLP, whose business today is mostly made up of coal-fired power plants, has promised to slash the amount of carbon dioxide that it releases into the atmosphere by decreasing its carbon intensity per kilowatt of electricity generated by more than three quarters. The CO_2 emissions intensity of its generating portfolio is targeted to fall from the 2007 level of 840 grams of CO_2 for every kilowatt of electricity to 600 grams by 2020 and 200 grams by 2050. By 2012, the company was at a level of 770 grams and proclaimed itself on track to meet the intermediate 2020 target.[2] For CLP to transform itself from an old-line coal-burning power company into a decarbonized utility is not simply a matter of slapping some solar cells on the side of a building or putting up a few wind farms. This is a planned transition to low-carbon power production that will take decades. It is a

British Prime Minister Margaret Thatcher speaks at the 1982 opening ceremony of China Light & Power's Castle Peak Power Plant, located in the northwestern area of Hong Kong. The $2.6 billion that China Light invested in Castle Peak was largely spent on British equipment, making it one of the largest export orders that British industry had ever received. Photo credit: CLP Holdings

transition that is measured not in dramatic, earth-shattering steps but in small incremental ones.

Kadoorie and his team have laid out a road map for a low-carbon future that will see 20 percent of CLP's power come from renewable energy (hydroelectric, wind, solar, and biomass) and 30 percent from non-carbon-emitting sources, including nuclear in addition to renewables, by 2020. Although these goals may sound ambitious, the company is well on its way to meeting them—in fact, the 2020 goal for power from non-carbon-emitting sources was raised from 20 percent to 30 percent after the original target was met ten years ahead of schedule.[3] By 2012, the company had taken the share of its electricity generated by renewable energy from less than 1 percent in 2004 to 20 percent with a series of investments ranging from wind farms in India to a solar farm in Thailand to a cotton-waste-burning biomass facility in China. By the next year, 2013, it was the largest foreign investor in renewable energy in China as well as the largest wind energy developer in India.[4] From 2007 through 2012, it invested almost $3 billion in renewable energy projects.[5]

No major utility in the world has voluntarily made a pledge like CLP's. The company's goal is in line with scientific estimates that developed countries need to cut their greenhouse gas emissions by 80 percent in order to stabilize carbon at 450 parts per million and prevent a dangerous acceleration in global warming. It is also a bold target, for the company doesn't know if the technology or government policies needed to reach this goal will be in place.

Kadoorie didn't have to choose to do this. Neither China nor Hong Kong, where the company has most of its power plants, is bound by the Kyoto Protocol to cut greenhouse gas emissions. But CLP hasn't done this only because it is the right thing to do. It has made this decision because its owners manage for the long term. Kadoorie and his staff don't profess to be smart enough to fully understand climate change, and no one knows precisely what sort of cuts in carbon will need to be made by midcentury. But they are convinced that climate change is real. They know that the rules of the game will change. And they are betting their company's future that staying ahead of its competitors, and of public opinion, will pay dividends when carbon emissions are more tightly regulated, whether through taxes or a trading scheme.

CLP is conscious that its business depends on a social license to operate and that, sooner or later, its carbon emissions will be regulated and capped. Better to stay in front of the politicians and the public than behind.

To hear Kadoorie tell it, this is the sort of thinking that comes naturally to a family running a company that is over one hundred years old and planning for its next century. CLP's former CEO, Andrew Brandler, told me that Kadoorie challenged him to think in fifty- or even one-hundred-year cycles. It is an unimaginable length of time for many Western CEOs. But it is the time frame that is necessary for a company that wants to shift $20 billion of assets from old-line fossil fuel sources like coal to low-carbon energy in a way that preserves the economic viability of the company. It gives new meaning to the notion of sustainable investment, and it is fitting for a family whose motto is "Adhere and prosper."

~

The Kadoorie office in Hong Kong's Central district doesn't look like it's home to any sort of green business or to someone intent on moving billions of dollars in coal-burning assets into low-carbon power sources. Only the carpet is green—and thick, in an old-fashioned, 1970s sort of way. A model of a helicopter landing pad that Kadoorie hopes to see built in downtown Hong Kong is in the lobby; helicopters are another of his passions. A Nepalese Gurkha guards the Kadoorie sanctum.

Kadoorie comes to greet me in the waiting area. He is short and energetic, with an unmistakable resemblance to the photographs I have seen of both his father and his grandfather. Best known to the public for his luxury Peninsula hotels, he typifies the family's penchant for a curious mixture of grand displays of wealth combined with a keen sense of social responsibility and a willingness to make business decisions looking ahead decades or even centuries. Hong Kong's reclamation has shrunk the view, but from his corner office, he can still see across the harbor to Kowloon, where CLP got its start more than a century ago as China Light & Power.

"The family has always thought long-term," says Kadoorie.

> The advantage of thinking long term is that you don't need to make decisions with respect to your capital involvement in the light of having to see a return in three months' time. . . . You are not chasing yourself. . . . If you wake up in the morning and you are owned by Post Krispies or someone different you are constantly changing your philosophies. You would be unable to instill in your management the confidence to take certain projects forward because they would be too short-term.
>
> That has been the philosophy of the family in virtually everything. If you are going to take the trouble to review or look at a scheme, you might as well see that the life of a scheme is going to be as long as possible. It eliminates a number of projects that are simply not sustainable, not long-term, and therefore are not on your horizon.

During most of its first one hundred years in business, China Light & Power was a rather ordinary electricity company—or at least as ordinary as events like the Japanese invasion of Hong Kong in 1941 and a communist revolution just across the border in 1949 could allow. When the company was founded in 1901, Hong Kong was a British colony with a population of 280,000. The first generating station was built on an isolated strip of land that was on the Kowloon side of the colony's spectacular natural harbor. The oil-fired plant had a capacity of just three 75 kW generating units, and peak power demand that year was one tenth of a megawatt—equivalent to having a thousand 100 W lightbulbs turned on at once. That wouldn't be enough to power a small Hong Kong apartment building today. The company had fewer than two hundred customers, and rather than using electricity meters, it charged on the basis of how many lightbulbs a customer owned.[6]

Elly Kadoorie was an extraordinary entrepreneur, and China Light & Power was only one of many investments. Some of his earliest investments were in hotels, and today the family is known internationally for its Peninsula hotels, whose shares have traded on Hong Kong's stock exchange (as Hong Kong and Shanghai Hotels) since 1920. The Peninsula in the waterfront district of Tsim Sha Tsui in Kowloon is the Kadoories' flagship hotel. It opened in the 1920s with a sweeping view of the harbor and was only a few steps from the terminus of the Kowloon Canton Railway, which allowed travelers to journey as far as Berlin and Paris, via Beijing and the trans-Siberian railway.

It was in the Peninsula Hotel that Hong Kong Governor Mark Young surrendered to the Japanese military on Christmas Day 1941, and the Japanese used the hotel as their first headquarters in Hong Kong. The war was devastating not just to the business but to the family as well; Elly Kadoorie, under house arrest by the Japanese in his vast Marble Hall estate in Shanghai, died of prostate cancer in 1944. Michael Kadoorie's father, Lawrence, was interned twice by the Japanese. Sent first to a prison camp on Hong Kong Island, he went to Shanghai upon his release and was once again imprisoned by the Japanese.

When the family returned to Hong Kong from Shanghai in 1945, its businesses were in ruins. China Light executives had deliberately destroyed key parts of the newly opened Hok Un Power Plant as Hong Kong was near surrender to the invading Japanese. Although the plant was partially repaired by the Japanese, a fuel shortage forced power cuts in 1944. Near the end of the war, as coal supplies dwindled, the Japanese-run plant burned rice husks and pine trees for fuel—an early, if unplanned, experiment in biomass power generation. The turmoil wasn't over.

In 1949, Mao Zedong and his Communist fighters swept to power in China. In the wake of the Communist takeover, the Kadoories lost all their assets in the country. The United Nations embargo of China during the Korean War cut Hong Kong's trade ties with China and further challenged the city's growth.

But the Kadoories were incurable optimists and focused on rebuilding from the ruins and destruction of war. During the Japanese occupation, the colony's population dropped from 1.8 million to 600,000, but in the early 1950s, the population surged to 2.6 million as refugees fleeing Communist China streamed into the city.[7]

What followed were China Light's heroic years in Hong Kong, as the company struggled to keep up with growth. During this time, the company generated ever-increasing electricity—and profits—from oil- and coal-fired power plants. As Hong Kong expanded, so, too, did China Light. From 1951 to 1981, its electricity generation capacity increased more than fiftyfold, from 50 MW to 2,656 MW.[8] By the end of the century, it had more than 10,000 MW of generating capacity.

But in the early 1990s, China Light, which was now CLP Holdings, suddenly faced a crisis: growth in electricity demand stopped dead as Hong Kong's manufacturing base withered. The city, once dubbed one of Asia's "Four Tigers," suddenly saw its manufacturers move across the border to newly opened China, where wages were cheap and workers were plentiful. It was up to Michael Kadoorie, newly in charge following his father's death in 1993 and his uncle's in 1995, to chart a way forward for the company. Adding to the challenge was the uncertainty of what would happen after July 1, 1997, when Hong Kong would become part of China. Should the company stay a midsized utility in a city whose high-growth years were behind it? Or should it venture abroad? "Every generation, every decade there have been different challenges," says Kadoorie. "What is very important is that when you see change on the horizon you actually take these challenges and you don't stand still."

Some members of the board wanted to take a conservative approach, but standing still was not for Kadoorie. He didn't want his legacy to be running down a century-old family asset. Asia was flush with capital, and the privatization of state-owned utility monopolies was fashionable. It no longer made sense to have state-owned power companies doing a job that the private sector could do more quickly and more cheaply. Recognizing the opportunity, he won the board over to the idea of a regional growth strategy.

Taiwan was among countries welcoming foreign investors in the power sector, and CLP won a contract in Taiwan's first round of bidding. With the Taiwan Cement Co. and Japan's Mitsubishi, the company built the 1,320 MW coal-fired Ho Ping Power Station on the island's northeastern coast. It bought a large coal-fired power producer in Australia and a significant stake in Thailand's dominant electricity producer as well as a stake in a Philippine power plant. What all these investments had in common was coal.

~

Billions of dollars of CLP's investments are tied up in facilities like coal-fired power plants. They physically can't be moved. They can't be traded like stocks or a currency. They cost a lot of money, and they last for decades.

Today, CLP is still mostly an old-line coal company. It has ten coal-fired plants in China. In 2013, CLP released 44 million metric tons of carbon dioxide into the atmosphere—an improvement over the nearly 50 million metric tons it emitted in 2009 but nonetheless about the same as all of Switzerland. That same year, CLP released 50,500 metric tons of sulfur dioxide and 50,200 metric tons of nitrogen oxides.

Richard Lancaster, who ran CLP's Hong Kong operations until he succeeded Andrew Brandler as CEO in late 2013, took me to see the giant Castle Peak Power Plant one afternoon in September 2010. Located on the northwestern coast of Hong Kong, the massive plant is still in a relatively isolated part of the territory.

The first and most obvious point is that coal plants are almost unimaginably large. Castle Peak is larger than the size of seventy-five soccer fields. Its two towering power stacks are each two thirds the height of the Empire State Building. When the plant is going full blast, it burns twenty-four tons

Jinchang Solar in Gansu is CLP's first solar project in mainland China. It began operation in July 2013 and has a capacity of 85 MW. Photo credit: CLP Holdings

of coal a minute—for a sense of scale, think of twenty-nine classic Volkswagen Beetles shoveled into the massive boilers every minute—that's one every two seconds—hour after hour.

The original $2.6 billion that CLP invested in Castle Peak, much of it on generating equipment built in Britain, made it one of the largest export orders that British industry had ever received. British Prime Minister Margaret Thatcher thought Castle Peak was important enough that she presided over its 1982 opening. Even today, its 4,108 MW of generating capacity make it one of the world's largest coal-fired plants.[9]

Castle Peak is a wonderful monument to industrial efficiency and might. It is also, sadly, the single largest source of emissions in Hong Kong. CLP disputes that it is a major cause of the pollutants that Hong Kongers breathe, citing its remote location and the prevailing winds as well as the extraordinary amount of pollution produced across the border. And the company notes that Castle Peak sticks out in part because so much of the polluting industry moved across the border to Guangdong Province in southern China. But the fact remains that it spews out more particulates than any other single facility in the territory.

In Asia, most electricity is provided by coal-fired power plants like Castle Peak. Coal is cheap and plentiful, and its abundance poses a question for utilities like CLP. Most companies have chosen the quick and dirty answer, which is to burn what coal they can today and make modest investments in alternatives. Current executives figure that they will almost certainly be gone by the time the real crunch on climate change comes. The investments in alternatives that many utilities make are a sideshow, a combination of public relations spin, political insurance, and development of technical expertise for the company.

CLP has taken a different approach. It has laid out a series of milestones in order to fulfill its promise of reducing its carbon intensity by three quarters by the middle of the century. How it decided to make that promise and how it intends to keep it set it apart from other power companies.

~

When Andrew Brandler joined CLP as the new CEO in May 2000, the company was facing a complex dilemma, one that he (and most other executives) had little experience with, despite his impressive credentials. He'd done his undergraduate studies at Cambridge and later earned an MBA from Harvard. Before joining the company, he ran the Asian arm of the venerable British investment bank Schroders. Personable, articulate, and intelligent, Brandler had specialized in nurturing relationships with energy companies such as CLP. While with Schroders, he had started to see carbon-related restrictions and clauses popping up in loan agreements; by the time he joined CLP, the issue of greenhouse gas emissions and pollution was starting to emerge as an important concern.

The extraordinary growth of Hong Kong and the surrounding Pearl River Delta had visibly blackened the city's air during the 1990s. CLP's boast that its Castle Peak Power Plant was one of the world's largest coal-burning plants was no longer a source of pride to people who had to breathe the soot from its smokestacks. Public anger about Hong Kong's increasingly dirty air was growing, and with Castle Peak the single most important source of particulate pollution in Hong Kong, the company was one of the top targets of public wrath. Greenpeace hammered at the company with high-profile protests.

Complicating matters for Brandler was CLP's drive for regional expansion, fueled by investments in coal. It had recently bought the Yallourn Power Plant and its supporting mine in Australia, which even today is one of the country's largest open-pit mines. The power station was still producing one fifth of the electricity for the populous state of Victoria, but the mine's brown coal was very carbon-intensive. Incredible as it seems today, CLP did not consider Yallourn's greenhouse gas emissions when it bought the plant and the mine in December 2000. But the Kyoto Protocol and other negotiations soon made it apparent that the economics of investing in coal-fired power plants were going to change and that there would be limits of some sort on greenhouse gas emissions. Utilities suddenly needed to factor this new business risk into their thinking.

Brandler remembers the strategic dilemma that faced the company when he took over. "We are in Asia. We have a mandate to grow the business. But if we do nothing but coal we will be crucified—maybe not today but someday. How do we move forward with a coherent long term strategy?"

To help find an answer to the puzzle, Brandler turned to Peter Greenwood, his iconoclastic deputy. Greenwood's background was in British military intelligence, but he'd spent much of his career as a corporate lawyer. Greenwood first moved to Hong Kong in the mid-1980s as part of the team that built Hong Kong's new airport—one of the world's largest engineering projects. He joined CLP a decade later.

"The starting point was a confluence of events," says Greenwood of CLP's transition away from its reliance on coal. "First, there was growing concern about poor air quality in Hong Kong. Second was the ignorance about the difference between air quality and climate change." Indeed, the soot in Hong Kong has little to do with global warming. The particulates that turned the city's air a semipermanent and disconcerting gray-brown are made up of sulfur dioxide and nitrogen oxides—pollutants, not greenhouse gases such as carbon dioxide, an odorless, tasteless, and (in the right concentration) generally benign part of our atmosphere.

As the twin issues of air quality and climate change rose on the list of public concerns, Brandler and Greenwood pondered what to do. CLP risked being caught up in political forces of a sort that it had never faced before. "We might have just dealt with air quality first and climate change later," says Greenwood. However, because the company worried that it was losing the public relations battle on air pollution and was aware that the issue of global warming would soon be similarly contentious, it decided to tackle both problems at once. Jane Lau, the head of the public relations team at the time, scrambled over just a few weeks in 2004 to write a pledge to grow the company's renewable energy business by 2010. Titled the *Manifesto on Air Quality & Climate Change*, it was the first of a series of documents in which the company went public on its thinking.

A good deal of internal controversy dogged Lau's project. As Greenwood remembers, there was a debate about "whether we should have a target at all." After all, CLP couldn't reach the renewables target on its own—government policy would be key. And if the company did decide to have a target, should it be ambitious or easy? Should it be published? The company knew that some companies had targets but didn't publish them. It worried that publicizing the target could be commercially damaging, driving up the prices demanded by potential partners or equipment suppliers who knew that the company had an artificial renewable-energy target to meet. After vigorous internal debate, the company publicly adopted the most aggressive of its targets: to have 5 percent of its power come from renewable energy, excluding nuclear, by 2010.

Having set its goal, CLP was now faced with the much more challenging task of trying to meet it. Its staff was expert at running power companies and building electricity grids, but the company wasn't set up to address an issue like the future of energy or climate change. So the company did what it had done for generations: think very long-term and hire expertise from outside.

CLP's search led it to Dr. Gail Kendall, who had been a professor at the Massachusetts Institute of Technology. "I'm not technical," says Michael

Kadoorie. "'I have an interest, obviously, in technology, but none of us are really knowledgeable in these areas. To be sure that you were being kept up to date as a layman, and I say that as a layman, I felt that [Kendall and her team] would bring the various disciplines and the various new thinking at least to the horizon of China Light. And then if there was any particular trend that needed any further thought we had enough people within China Light who could address it."

Being an American woman in a company run largely by British and Hong Kong Chinese executives and engineers certainly set Kendall apart from the crowd. Intense and cerebral, she had a high-level awareness of emissions issues that went far beyond the immediate concerns of Hong Kong's deteriorating air quality—and, crucially, she had the backing of Kadoorie and Brandler. Kendall "had the freedom and flexibility to look forward," says Greenwood. "She was not part of day-to-day management but was brought in to bring a forward-looking vision."

Kendall's arrival coincided with the creation of the CLP Research Institute, a stand-alone unit that was charged with building renewable-energy projects. The institute was set up to provide an entrepreneurial culture that would let the new ideas flourish. Jeanne Ng, one of the unit's most prominent members, was a Hong Kong–Canadian PhD and self-described "green loony." She originally studied toxicology and worked for the Toronto police for a summer. She had planned on a career as a CSI-style pathologist before her more traditional parents decided that police work wasn't the job they wanted for their daughter. Hiring people like Kendall and Ng and setting up a unit where entrepreneurial thinking could be incubated proved key to CLP's breakout in renewables. So, too, did the growing commercialization of renewable energy, as wind turbines were becoming more reliable and dropping in price. The company not only met its ambitious 5 percent goal but far exceeded it. It finished 2010 with 15 percent of its capacity in renewable energy, three times the target and about fifteen times the level in 2004. It has since set more long-term targets, looking forward into the next half-century.

~

A string of forty-six towering wind turbines dots a sweeping semicircle of coastline along China's massive Bohai Bay. Standing white and silent as they whir, their 231-foot-high wind masts dwarf the fishermen who gather seaweed for their dinner tables. To the east of this northern tip of Shandong

Province, three hundred miles across the Yellow Sea, lies the South Korean capital of Seoul, closer than Beijing.

Not far from the wind farm is Weihai, a city of 2.5 million people, where swathes of modern apartment blocks are swallowing fields and farmhouses. A new expressway lined with lights ties the city to the airport and its recently built terminal. Downtown, a massive glass-faced convention and exhibition center sprawls along the harbor. As almost everywhere in China, Weihai is seeing new apartments and offices sprouting up with a speed that would seem fantastical in Europe, the United States, or Japan.

Shandong, already one of China's richest provinces, exhibits economic growth that looks like it will keep racing along for some time. Fertile fields, a rich trading tradition on the coast, and booming factories are coupled with an optimism and a desire to get ahead that are almost palpable. The relentless growth that is taking place in this province and in cities like Weihai means a hunger for resources. Weihai needs power to keep the convention center and the airport heated and cooled, the road to the airport lit, and its people living in the comfort that comes with a modern life. So do all of China's other cities.

The wind turbines that line the coast are a small part of the solution to China's energy problem. The country has the world's largest base of wind power, and CLP is the largest private-sector investor in China's wind power industry. Among its many investments in China, it owns a 49 percent share of the Weihai wind farm.

Wind is a wonderful renewable energy. But it poses problems for anyone trying to run a power network. Besides their inherent unpredictability, winds are not as strong nearer the equator, so for much of tropical Asia wind power is not a good alternative. It's also expensive to transmit power from the often remote regions where the wind power is generated to the more populated places where people need the power. All that said, wind is a fairly cost-efficient renewable power source—certainly, for now, more so than solar. But in rough terms, in China it is still twice as expensive as coal power. So CLP needs to focus on other renewables as well if it's to have any hope of meeting its goals.

~

Another part of the solution to the puzzle of how CLP can get green was in a little bottle that I received as a souvenir a few weeks before I met Michael Kadoorie. The oversized snuff bottle has a small power plant painted on the glass. Inside are four small twigs, each two inches long and the thickest no more than half an inch in diameter. The twigs are cotton stems, the waste from the bushes left after the cotton is harvested. Typically, these agricultural leftovers are burned by farmers, just as American suburbanites once burned autumn leaves. The damp, smoky fires are choking, causing respiratory problems for those who live nearby and posing a hazard for everyone from airline pilots to car and truck drivers.

Yesterday's waste is today's biofuel. CLP teamed up with a factory in the town of Boxing, Shandong, to build a plant to produce heat and power with the cotton waste. CLP has demonstrated that a hunger for energy can, with creative business practices, at the same time cut pollution.

The Boxing Power Plant is sophisticated and sizable. On a visit to the site, I saw massive mounds of what on first glance look like gargantuan bread loaves. The loaves dwarf our group: they are thirty-three feet high, forty-five feet wide and a staggering two hundred feet long. Measured from end to end, each of these stretches nearly two thirds the length of a football field. These hills are symmetrical and carefully sculpted piles of mulched cotton waste. They are evenly spaced, for there is always the danger of fire in these forty thousand tons of fuel.

The plant takes the cotton waste and produces power. This $30 million plant, which at the time I visited was majority-owned by CLP but was subsequently sold, plays three important roles. It cuts the pollution produced by farmers who would otherwise burn the waste in their fields, reduces by just a little bit the need to burn coal, and provides local employment. In November, after the cotton harvest, lines of trucks queue up outside the plant to deposit the leftover cotton stalks and husks. The waste is mulched in a primitive process that uses an oversized wood chipper (reminiscent of those used by contractors and towns in suburban America) before being piled into hills. The waste is fed onto a conveyor belt that takes it into the power plant where it's burned in an oversized furnace. The high burning temperatures of biomass plants mean more complete combustion and less pollution than leaving farmers to burn the waste in their fields. The fire heats water, which produces steam; the steam drives turbines; and the turbines in turn produce electricity. The steam-driven turbines are the same as those used in conventional power plants. The fire? Except for scale and efficiency, it's not much different from the sort of wood-burning stove found in many homes.

Boxing is a small plant by the standards of the massive coal-, gas-, and nuclear-powered plants that CLP operates. But it produces power that's used locally, reducing the electricity that's lost in transmission over long distances. The electricity cuts the need for oil- or coal-fired power and at the same time reduces the uncontrolled burning of agricultural waste by farmers.

The biomass plant also produces steam, which is used by a paper plant next door, cutting down on that factory's need to burn coal or oil. About one thousand local farmers earn a combined ¥40 million(a bit over $6 million) a year in extra income by selling this waste to factories. They no longer have to burn the cotton waste with fires whose blanket of smoke has choked those in its path and prompted car and plane accidents.

Each year the Boxing facility consumes 150,000 tons of cotton waste, emitting about half the greenhouse gases as would a similar amount of coal. The venture sells 130 million kilowatt hours of electricity a year and 130,000 tons of steam. The electricity goes into the regional power grid and is used by five nearby paper mills and one chemical plant.

All this sounds impressive, and it is—but on a small scale. All this work produces a modest 15 MW of power. A nearby coal station in which CLP has a stake produces eighty-four times as much. It is hard to generate much power with an individual biomass plant, and Boxing's managers scour the countryside for many miles to get enough fuel. Top CLP managers spent a disproportionate amount of time on what is a very small energy source. Coal has succeeded for a reason.

~

CLP started considering nuclear power long before it began its push toward greener forms of energy. In 1980, Michael Kadoorie's father, Lawrence, presented the Chinese with a five-volume feasibility study on building a nuclear power plant in Guangdong. Nuclear power already had been studied and rejected for Hong Kong, but Lawrence Kadoorie knew that a country of China's size would need every kind of power it could get if Deng Xiaoping's aspirations for Chinese economic growth were to be realized.

This was a revolutionary idea. Mao Zedong had died only four years earlier, and although Deng Xiaoping had said that China was "open for business," there were many doubters. Lawrence Kadoorie was not one of them. In 1985, he met Deng Xiaoping and agreed to invest in a nuclear power plant at Daya Bay, just over the border from Hong Kong. This was both China's first commercial nuclear reactor and by far the largest single foreign investment in China since the 1949 revolution. China Light agreed to buy 70 percent of the plant's electricity, providing the promise of much-needed foreign exchange at a time when China was waking from its economic slumber.[10]

The China of 1985 was poor. Average annual incomes in cities were about $100 and half that in the countryside.[11] Although the country had exploded a nuclear device back in 1964 and had nuclear weapons, it had no commercial nuclear power plants.[12] Its scientific ranks had been decimated by the Cultural Revolution. Universities had been shut during the height of the turmoil, with China's scientists, technicians, and engineers packed off to the countryside to do menial labor and supposedly learn from the masses.

Although the intellectuals and elite made their way back to the cities in the 1970s, China was still in a shambles when Lawrence Kadoorie made his bid to help China build a nuclear power industry. Despite intense opposition in Hong Kong, including a petition against the plant signed by one million people, the Daya Bay Nuclear Power Plant was built. Construction spanned the dark days around the June 4, 1989 Tiananmen killings, when China's economic reforms lurched, shuddered, and seemed to stop. Western countries put on a variety of trade sanctions, China's leaders were not welcome abroad, and a million people marched in the streets in Hong Kong protesting the Tiananmen violence. With the British colony slated to revert to Chinese sovereignty in 1997, the future for a multibillion-dollar nuclear power plant in China did not look promising.

Fears about the safety of China's nuclear program were fueled when it became known that the plant's containment vessel had some apparent defects in its construction. The containment vessel is the last line of defense in the event of a meltdown or catastrophic accident. It is the iconic dome of nuclear imagery, lined on the inside with steel designed to keep radiation inside the building. It is built to extremely rigorous specifications so it is able to stand firm against a jet plane flying into it or a massive earthquake. In the case of the Daya Bay plant, some reinforcing bars that were meant to be included in the vessel were mistakenly left out. In a basic human error, a quantity surveyor had miscounted the number of rods needed. If the contractors couldn't get the number of rebars right, many people wondered, how could China be counted on to run the nuclear plant safely?

Today, a pillar of shame stands near a viewing platform overlooking the Daya Bay plant and the neighboring Ling Ao nuclear complex. This is one of the improperly constructed concrete piers for the containment vessel, a reminder of the need to avoid mistakes. In the aftermath of the incident, which was discovered while construction was under way, modifications were made to buttress the containment vessel. Although CLP takes credit for making the incident public, it's a sign of China's increasing confidence and transparency that the pillar is in such a prominent place and is pointed out to visitors as a reminder that there is little margin for error at a nuclear plant.

The Daya Bay venture was a seminal one for China, and the country is now embarking on one of the world's most ambitious nuclear programs. Plans call for a sixfold increase in nuclear capacity over the next decade, with at least 58 GW of nuclear power by 2020. CLP wants to be a big part of this and is negotiating to deepen its involvement in China's nuclear industry. Achieving this expansion will be difficult; training enough people and manufacturing enough plants will be a challenge. But someday, in the not-too-distant future, Brandler expects the export of China's nuclear plants to Europe or the United States. "I have no doubt that China will build reactors in the United States—it sounds far-fetched now, but it will happen," CLP's CEO told me in 2011. In fact, in June 2014 Chinese Premier Li Keqiang signed an agreement with British Prime Minister David Cameron to build Chinese nuclear reactors in the United Kingdom. Even with China's ambitious plans for its domestic nuclear industry, plans that will see one of the most rapid build-outs in history, nuclear power in China will account for a paltry 5 percent of power in 2020 (compared with, for instance, 75 percent in France).

The great drawback to nuclear power, besides a catastrophic accident such as Chernobyl or Fukushima, is what to do with the waste. A thirteen-minute introductory video to Daya Bay says that the high-level waste is sent to northeast China for reprocessing. What it does not say is that the material is being stored. No reprocessing has taken place. The Daya Bay Nuclear Power Company has found a permanent storage site for its low-level waste ten kilometers away and is awaiting official approval for this planned solution. Until then, the low-level waste remains in pools at Daya Bay. Here China, for all its authoritarian ability to get things done, faces many of the same challenges as other countries, although because its nuclear program is so small, it has less of a problem than countries like the United States.

Nonetheless, it is telling that even a country with a centralized, authoritarian, and highly technically competent government is grappling with same waste disposal issues as democratic countries. In answer to a question, Daya Bay executive Steven Lau bristled: "Why isn't America doing it? . . . What is happening with Yucca Mountain?" (He is referring to the proposed but long-delayed U.S. national storage site.) Lau said that many technological and financial issues remain to be ironed out, but he allowed that "you can say that it is political." The nuclear industry is secretive everywhere. Add China's reflexive secrecy, and it's no wonder that people in Hong Kong are mistrustful. The issue of waste disposal simply won't go away. To pretend that the problem has been solved when it hasn't undermines credibility.

Continuing concerns about safety remain in Hong Kong. In June 2010, it was learned that a few weeks earlier, on May 23, the water in one of the reactors had shown a slightly elevated level of radiation. An outcry followed news of the incident. The territory was in the midst of a debate about a political reform bill, and the debate over nuclear power's safety called into question how transparent and reliable China was as a partner. Much as CLP management preaches openness, some of its executives have a defensive attitude in dealing with critics. During our visit to Daya Bay, Lau characterized the reports on the incidents as "rumors," although in fact details had been posted on the official website. There is in so much of the nuclear industry a curious mixture of bravado and evasion—the assertions that another Chernobyl or Three Mile Island or, now, Fukushima, simply could not happen; claims that the technology is virtually foolproof and essentially perfect; but also the admission that, as in the case of fuel rods or even Fukushima, we shouldn't, of course, expect perfection.

These caveats notwithstanding, CLP expects that nuclear will play a big role in the transition to a low-carbon future. Building on its experience with Daya Bay, CLP plans to make significant new investments in China's nuclear program. In July 2010, it signed an agreement for a 17 percent share in a new nuclear plant being built at Yangjiang, which like Daya Bay is in the southern coastal province of Guangdong, which is adjacent to Hong Kong. However, in the wake of the Fukushima disaster, China slowed down its nuclear program and the two sides did not sign a final agreement. In September 2013 CLP announced that negotiations had been called off.[13]

Certainly, Michael Kadoorie is keen to keep investing in nuclear, especially in China:

> We hope to increase our nuclear profile, which is in fact today probably the cleanest sustainable energy source for base load. We've had a long ongoing and profitable relationship with our mainland counterparts, which they and ourselves wish to extend and that is in fact what we are negotiating. We feel very fortunate that we have been welcomed [to China's nuclear program]. What is your base load going

> to be? Is it going to be wind, which never comes? Is it going to be hydro, which you can't get involved with to a great extent because the opportunities aren't there? Is it going to be sunlight? No. Today, this is where we are.

~

Wind power and solar power weren't part of CLP's portfolio of electricity-generating businesses as recently as 2000. Yet by 2013, renewables made up 20 percent of the power the company sells around the region. Long reliant on coal power, it is generating electricity from the wind, the sun, and, for a time, even such unlikely sources as cotton waste. These are commercial projects. In some cases, they are small, but the total is significant. Already, as noted, the company is the largest investor in wind power in India (it has 1,000 MW of installed capacity) and the largest foreign investor in wind power in China.

CLP is the largest wind energy investor in India, with over 1,000 MW of wind capacity in the country as of mid-2014. Photo credit: CLP Holdings

Yet this is only the beginning. The bigger changes will come over the next thirty-five years. CLP has already said that it will no longer buy or build the dirtiest coal-fired plants, and it has started selling some of its coal plants. But when it comes to coal, there's really no such thing as clean. The company will have to make some hard choices to deal with its remaining coal-fired electricity plants.

CLP is going to need a government framework. There is only so much that a private company can do without clear rules of the game. Right now the company is groping its way toward a low-carbon future in an era of changing technology and with no international framework. It works with national governments, tailoring its investments to the prevailing policy environment. Its first major solar investment was possible only because of favorable regulatory policies in Thailand. Its large wind investments in India are economical possible only because of supportive regulatory policies at the state level.

For now, CLP promises not to build any more conventional coal-fired plants in Hong Kong or other developed economies where it owns assets, such as Australia.[14] That will mean more wind farms, more solar plants, and, perhaps most significantly, more nuclear power. The alternative, business as usual, is a grim option. In 2006, the year before the company made its decarbonization pledge, its coal-fired power station in Hong Kong alone emitted eighteen million tons of CO_2 emissions. Thanks to its investments in cleaner fuels—nuclear and natural gas—as well as in emissions control technology at its Castle Peak coal-fired plant, its total Hong Kong emissions fell more than 80 percent from 1990 until 2012, according to Michael Kadoorie, despite its producing 81 percent more electricity.[15]

Thirty-five years sounds like a long time. But it is little more than the blink of an eye in the electricity business. Many of the power plants that are being designed today will still be in service in 2050. CLP's Yallourn plant in Australia started operations over forty years ago, in 1974, and still produces 8 percent of the entire country's electricity. The easiest course of action for electricity companies is to keep doing more of the same until forced to change. This sort of "if it ain't broke, don't fix it" approach makes superficial sense in a business where a new power plant can cost upward of $1 billion. The power industry, understandably, is not one that generally rewards out-of-the-box thinking. The goal for the past century of electrification has been simply to power more lights and televisions and air conditioners—and to make sure that the power never goes off.

With the consequences of climate change becoming more apparent, it is obvious that the old get-dirty, get-rich, get-clean model of economic development is broken. CLP is unusual in being an old company that is engaging in new thinking in carbon-intensive industries. That fundamental rethinking of its basic business took place in the early 2000s. We are now a decade into what is likely to be a fifty-year process of unfolding what it means to be a big power company in a low-carbon world. When we're in the middle of history being made, we don't know how it will turn out. Today, we do know that the company is a pioneer. Some pioneers prosper; other pioneers end up with arrows in their back.

"Adhere and prosper" has worked well as a Kadoorie family creed for 130 years. Michael Kadoorie is counting on it as a guiding principle for the next 100 years or more. For the world's sake, we should hope that he and his team succeed.

Conclusion: From Black to Green: *Asia's Challenge*

> We will resolutely declare war against pollution as we declared war against poverty.
>
> —CHINESE PREMIER LI KEQIANG

Asia's environmental emergency threatens personal and even national survival. Every year, air pollution causes some six million premature deaths around the region, and climate change and sea-level rise threaten to put some countries completely underwater. Lakes and rivers are undrinkable. Tropical forests are being stripped.

There is not one single model that all countries can follow to improve their situation; Asia is large and varied, and what works in Singapore might not work in Indonesia. Still, some basic principles are the same everywhere. Smart and strong policies, innovative and incentivized companies, and an engaged civil society are all crucial. Government, business, citizens—all have a role to play.

Responsibility for reducing carbon emissions and minimizing pollution and environmental degradation starts with individual governments. Governments need to set the rules with clear policies within which individual businesses and ordinary citizens can act. One important policy change that governments can undertake is the elimination of subsidies that encourage the use of fossil fuel, which total more than $1.9 trillion annually. Prices can have a powerful influence on behavior; users of fossil fuels should be forced the pay the real price for their choice, a price that includes health costs as well as the cost of carbon emissions. In most countries, both the right to emit carbon and the right to use water are dramatically undervalued. By changing this systemic underpricing and using prices as much as possible to influence behavior, governments can force companies and individuals to make their own decisions on how to use these precious resources more efficiently.

This book has focused on policies that are aligned with market forces. Policies that fight the market, rather than embracing it, are more likely to fail. Appreciating the increasingly favorable economics of renewable energy and environmentally sustainable policies is important to understanding that the greening of Asia is an economically sustainable goal. It would be naïve to think that price alone will be enough, but it is a bedrock principle. Eliminating subsidies and correcting the mispricing of valuable resources are among a number of policy interventions that governments can use to promote sustainability; others include progressively tighter standards for building, vehicle, and appliance efficiency; restrictions on carbon and pollutant emissions; and urban restrictions on vehicles, especially traditional gasoline and diesel cars and trucks. Concerted government-industry-university R&D partnerships can be powerful.

Businesses have a key and largely underappreciated role to play in the transition to a greener Asia. Companies have financial resources that allow them to make investments, often investments whose useful life extends many decades, and they can adopt technology more quickly and often more effectively than governments. They are flexible, changing course as needed to meet the demands of their customers and other stakeholders—workers, shareholders, civil society, and government. Above all, good companies have human resources in the form of smart, capable, and highly motivated employees who have a freedom to act that the public sector cannot—and should not—match. Business, in short, tries to get the job done as efficiently as possible. Business needs to be given the incentives—both rewards and punishments—to ensure that it acts correctly; governments and civil society need to establish these incentives.

More broadly, in this age of hypervigilant social media, companies as a whole must be increasingly responsive to the broader community, or they risk losing their social license to operate. In extreme cases, that can mean companies actually lose their assets, as some Taiwanese plantation owners in Sumatra, Indonesia, discovered in the 1998 political upheaval when nearby villagers simply took back land that they felt had been stolen from them. In less extreme cases, such as the melamine milk adulteration scandal in China, customers simply desert companies. From the effectiveness of Greenpeace and Western activists in changing forestry practices to the campaign against eating shark's fin, which has been heeded by such giants as Hong Kong Disneyland and Cathay Pacific, it's clear that protestors and organized consumers have the power to push for sustainability.

Governments, businesses, and civil society will, of course, be most effective when they work in partnership. That truism is especially apt in Asia, where consensus is prized as a political virtue. The best hope of mitigating an ongoing environmental nightmare in Asia—and it is a hope, not a certainty—lies in strong government policies coupled with business-led innovation and investment, backed by support from society as a whole. Of course, in every success story, the balance of these three forces differs.

In Singapore, for instance, the government has led the response to resource and environmental challenges. Key political leaders, influenced by strategic, economic, and environmental concerns, implemented new policies—first on water, then on green buildings, and finally on the fabric of the city. They built consensus from the top down, through ministry ranks and down to the grassroots level, using media, religious leaders, and a range of civic organizations to educate and engage ordinary citizens. Government money seeded a number of initiatives, spawning successful partnerships among universities, research institutes, and both foreign and local businesses. In Singapore's case, extensive government efforts have spurred corporate and societal support.

In Hong Kong, support from businesses wasn't quite so forthcoming. The city has a powerful business sector that has successfully fought many efforts to legislate environmental changes, and its government has been cautious about pushing through green laws. But when businesses have taken the lead, change has come. In 2012, after a group of shipowners and shipping lines drew up the Fair Winds Charter (a voluntary commitment to use low-sulphur fuel), the government provided a subsidy to reduce the cost of using this more expensive fuel. The shipping leaders wanted all ships to use the same standards, and at their request, the government introduced legislation making the cleaner fuel mandatory in the world's fourth-busiest container port.

The Philippines provides another example of a successful government-business partnership—and it's likewise a partnership in which the business, Manila Water, has played the leading role. The company took over a service area where water supplies were erratic at best and absent at worst and where about two thirds of the water leaked or was stolen from the aging pipes. Today, 99 percent of the service area's population has reliable water, and stolen or leaked water has been cut to an impressive 12 percent. Manila Water has also worked with the government on larger sustainability initiatives, like managing watersheds and operating sewage treatment plants. It has done all this while producing stellar returns for its shareholders. The company still has a ways to go in winning over full popular support; Manila Water has been pilloried by critics who claim that it is profiting on a basic human need at the expense of poorer Filipinos.

A good example of the importance of civil society is seen in the enthusiasm with which the Japanese embraced energy conservation measures in the wake of the 2011 Fukushima disaster. This high degree of national cohesion was also reflected in the rapid adoption of policies designed to jump-start solar power. The ensuing stand-off between large sections of the populace and the ruling Liberal Democratic Party over nuclear power is likely to emerge as a test case for citizen involvement in energy policy.

China's vast size and its mixture of planned economy, powerful local officials, and ambitious companies make it a singular case. Government backing is there, as least in words; in his annual state of the nation address in March 2014, Premier Li Keqiang promised that the government would tackle air, water, and soil pollution, as well as make changes in "the way energy is consumed and produced," citing nuclear and renewable power, the development of smart grids, and the nurturing of green and low-carbon technology. If the past is any indication, speeches of this sort are more aspirational than literal. Still, that the Chinese premier devoted significant parts of his annual work address to the environment underscores the important directional shift under way. At the top, officials clearly know that environmental challenges pose a significant problem. Yet China, generally regarded as a strong state capable of enforcing its will, so far has been unable to effectively implement a range of what appear, at least on paper, to be strong environmental protection laws, ranging from air pollution to water to food safety to building energy efficiency, due to resistance from local officials and powerful state-owned enterprises. It remains to be seen if Li's March 2014 declaration of a "war on pollution" will succeed.

Although the Chinese government hasn't proven to be a reliable driver of sustainability initiatives, it has been able to support the rapid growth of renewable-energy companies, which parallels the growth of the corporate sector more broadly. Both solar and wind companies have grown rapidly, and Chinese manufacturers now are globally significant in both industries. Local government provided the tax breaks, inexpensive land, and bank loans that were pivotal in this rapid growth. This growth (and the country's policies) has forever changed the global economics of renewable energy, but it has also proved disastrous to many of the individual companies involved.

Indeed, many of China's policies cause more environmental harm than good. The country uses far more scarce energy and at a far higher environmental cost than is needed, reflecting incentives given to local officials to prioritize economic growth and employment. In addition, its electricity is priced inexpensively, in no sense reflecting the environmental externalities of an electrical grid that is 80 percent powered by coal. Oil, water, and a variety of other resources are similarly underpriced, reducing incentives to use them more efficiently.

As of yet, there has not been a large-scale popular movement in China to address its growing environmental woes. Japan and Singapore and, to a lesser degree, Taiwan and South Korea have all made environmental initiatives a matter of national pride. Local builders and architects vie for environmental efficiency. Companies strive to win awards for efficient, environmentally friendly operations. This sort of bottom-up pride in energy-efficiency initiatives is something that China would be wise to nurture. Absent real enthusiasm from ordinary shoppers, students, and workers in offices, factories, schools, and shops, green energy efforts will fail to achieve anywhere near their potential.

Citizens can protest and sometimes vote; governments can set the rules; companies can provide the innovation needed for cost-effective solutions to energy and environmental problems. The preceding chapters showed a few of the many hundreds of Asian companies that have the willingness, the money, and the know-how to act (in the appendix on "Companies to Watch" that ends this book, you'll find a list of many of these innovative businesses). They need strong signals from their governments. Political leadership would ensure that the region's companies use their impressive human and financial resources to nurture the growth of a greener Asia.

Spurred by Asia's governments and supported by its cash-rich banks and capital markets, leading companies will use opportunities presented by its environmental nightmare to change the global economic landscape and in some cases overtake Western competitors in areas like solar and wind power, environmentally efficient vehicles, and green infrastructure. Just as Asia's developed economies in Japan, Korea, Taiwan, Hong Kong, and Singapore adjusted to higher wages by improving productivity and relying on better education and more innovation, so Asia will find a way to profitably do more with less in an era of resource constraints.

The great wealth and low population density in the United States give it the luxury of time when deciding how to respond to environmental challenges. Asia's resource and population pressures leave its countries with little choice but to respond. Authoritarian or democratic, governments need to meet an ever-more-affluent citizenry's demands for cleaner water, cleaner air, and more livable cities.

For Asia as a region and for its people as citizens of countries from giant China to the water-threatened Maldives, the fight against climate change, pollution, and environmental degradation is more than a metaphorical fight. The outcomes of this struggle will largely determine the region's prospects for prosperity in the coming decades. Asia will have to win its cleanup fight if this is going to be the Asian century. Good environmental policies, motivated companies, and effective civil society groups are key to a greener, cleaner, and richer Asia.

Appendix

Companies to Watch

This list consists of snapshots of companies and a handful of public organizations, highlighting their environmental sustainability initiatives. It includes principally companies and organizations mentioned in the book and is in no sense intended to be encyclopedic. Instead, I hope that it will give a sense of the range of companies involved in various environment-related businesses.

Asia Pulp & Paper (APP) APP, Indonesia's largest paper company and the third-largest in the world, has long been a target of environmentalists angered by its forest management practices. It took a significant step in early 2013 when it pledged to halt deforestation in its own operations and those of its suppliers. The pledge came after NGO activists boycotted the company's customers, such as Mattel, Nestlé, and KFC, costing it sales. Later in 2013, the company secured a $1.8 billion loan from China Development Bank to build a $2.6 billion pulp mill, the largest in Indonesia.

Austindo Nusantara Jaya (ANJ) ANJ, an Indonesian plantation company, focuses on palm but is now moving into sago production. "Black liquor" waste powers a biofuel power plant at its Belitung plantation, and excess power produced there is sold to the state electricity grid.

Among the company's sustainable practices is building nests for owls, who then hunt rats, eliminating the need for rat poison. Sacred burial sites on plantations are preserved, and corridors for orangutans and other mammals and buffers along rivers are maintained by the company. This reduces the amount of land that can be used for planting but preserves biodiversity.

AYALA GROUP Ayala, one of the largest and oldest business groups in the Philippines, has businesses in nine segments ranging from banking (Bank of the Philippine Islands) to property, manufacturing, telecommunications (Globe Telecom), and water. Manila Water (see separate entry) has been widely praised for its success in improving water services. In 2012, Ayala Land had a hand in building 28 percent of the LEED-certified buildings in the country.

BYD A Shenzhen, China–based battery company, BYD is controlled by founder Wang Chuanfu. A visionary engineer, Wang first made his mark producing batteries for mobile phones and other consumer electronics products; in 2003, he bought a struggling Xi'an automaker, retooled it to make electric cars, and emerged as one of China's largest battery-powered vehicle manufacturers. The company has struggled to sell electric cars; its current focus is on public transit, including taxis and especially buses. It continues research into large-scale storage devices for solar and wind farms. MidAmerican Energy, a subsidiary of Berkshire Hathaway, bought 9.9 percent of the company for $230 million in 2008.

CANADIAN SOLAR Although solar manufacturer Canadian Solar was founded in Ontario, Canada, the majority of its manufacturing plants are in China, and the company is dominated by Chinese and Canadian-Chinese senior executives. Sales in 2013 were $1.7 billion, and its workforce totaled seven thousand people. It has sold over 6 GW of solar panels. The company's customers include China's Three Gorges New Energy, China's Golden Sun Program, and Saudi Aramco's KASPARC project in Saudi Arabia.

CHINA LONGYUAN POWER GROUP China Longyuan, the largest wind farm developer in China and the second largest in the world, was formed from the merger of state-owned energy concerns. It raised ¥17.7 billion ($2.5 billion) from its 2009 listing on the Hong Kong stock exchange. The group also has significant coal generating capacity as well as smaller interests in geothermal, solar, and biomass. Its stock price nearly doubled in 2013 and it finished the year with 12 GW of installed wind capacity. China Longyuan plans to develop offshore wind projects and is expanding abroad.

China Ming Yang Wind Power Group China Ming Yang is the third-largest Chinese wind turbine manufacturer by 2013 installations after Goldwind and Guodian United. The company collaborates with a German partner and licenses its technology for turbines. Ming Yang listed on the NYSE in 2010 but has struggled with cash flow, as its highly concentrated customer base (Ming Yang's top five customers—mainly Chinese state-owned enterprises—make up 90 percent of revenue) has been slow to pay its bills. Investors suffered as the share price fell from a high of $14 in October 2010 to under $2 in late 2013 as the company struggled to be profitable. Prospects improved in 2014 but significant challenges remain.

CLP A Hong Kong–based utility previously known as China Light & Power, CLP is controlled by the Kadoorie family. The family's other major asset is the Peninsula chain of ultraluxury hotels. CLP expanded beyond Hong Kong in the 1990s, mostly with coal-fired electrical plants, but it rethought its focus on coal in the early 2000s. In 2007, it announced that it would reduce its carbon-intensity 75 percent by 2050. It is now one of Asia's largest privately owned utilities with 2013 revenues of $13.6 billion. It is one of the most significant private investors in wind in China and India; it has a solar venture in Thailand as well as a significant stake in China's Daya Bay Nuclear Plant.

Dongfang Electric Corporation Based in the interior Chinese city of Chengdu, Dongfang Electric is a large, state-owned enterprise that makes power generation equipment for traditional plants (coal- and even oil-fired) as well as hydropower, wind power, and nuclear power plants. Over 80 percent of its new orders are for high-efficiency and clean-energy generation equipment (hydro, wind, nuclear, and gas-fired), in line with China's policy to shift its energy mix toward cleaner sources. The company has 40 percent of the Chinese market for hydro power generation equipment and is expanding abroad, with offices and projects in India, Pakistan, Indonesia, and Vietnam. The company had sales of ¥42.4 billion ($6.6 billion) in 2013.

Esquel One of the world's largest shirt makers, Hong Kong–based Esquel has a workforce of almost sixty thousand; it annually makes more than 100 million men's dress and polo shirts for the likes of Polo Ralph Lauren, Tommy Hilfiger, Nike, Hugo Boss, and Lacoste; annual revenues top $1.3 billion. The vertically integrated company buys cotton from

Xinjiang farmers (and helps finance their water-saving drip irrigation) and then gins, dyes, and weaves the cotton before making the shirts. Its environmental efforts focus on water and energy efficiency. Chairman Marjorie Yang, a graduate of MIT and Harvard, is the third generation of a textile family originally based in Shanghai; daughter Dee runs the company's PYE retail chain.

FIRST SOLAR First Solar is the United States's only significant solar-panel manufacturer, and it ranks among the world's largest. The Tempe, Arizona–based company underwent a wrenching restructuring in 2012, moving away from reliance on a subsidy-dependent German market; its new focus is on providing large ("utility-scale") solar farms to customers including MidAmerican Energy. The company uses thin-film technology rather than the more common crystalline silicon solar process. Thin-film is less efficient at converting sunlight to electricity but cheaper to manufacture. First Solar acquired General Electric's solar business in 2013; GE in return took a 1.8 percent stake in First Solar.

FUJI XEROX Fuji Xerox is a Tokyo-based joint venture between Fujifilm and the American company Xerox. It had 2013 revenues of ¥1.02 trillion ($11.9 billion) and has over forty-five thousand employees. The company reduced factory emissions 10 percent from 2002 to 2012 and aims to cut emissions by 30 percent from 2005 levels by 2020. It has embarked on a program, in partnership with its suppliers, to ingrain environmental initiatives throughout its supply chain. For consumers, it has also introduced a number of recycling and zero-landfill policies.

GOLDWIND See *Xinjiang Goldwind Science & Technology*

GREAT GIANT PINEAPPLE Great Giant Pineapple, based in Sumatra, Indonesia, boasts the world's largest single pineapple plantation, at thirty-three thousand hectares. Its total pineapple production trails only Dole and Del Monte, and it employs nineteen thousand people. The company hopes that a move toward carbon-neutral status will give it a competitive advantage among consumers, and its sustainability focus has led to better traceability and more control over the entire food chain. A biogas production plant provides energy and also cuts the land needed for effluent storage, freeing up more land for productive use. Leftover pineapple pulp is fed to the company's substantial herd of beef cows; their manure in turn fertilizes the pineapple fields.

HANERGY Originally a hydroelectric power producer (it controls dams with a rated capacity of 6 GW), Beijing-based Hanergy has taken advantage of solar industry difficulties by acquiring technology-rich Western manufacturing firms using the less common thin-film solar process. Since 2012, it has bought Germany's Solibro, the thin-film subsidiary of bankrupt Q-Cells (Q-Cells itself was bought by South Korea's Hanwha); MiaSolé, a Silicon Valley start-up; Tucson, Arizona–based Global Solar, and Alta Devices, a Sunnyvale, California thin-film producer specializing in low light conditions. The company's 2013 revenues totaled $420 million.

HANWHA SOLARONE Originally founded as Solarfun in 2004 by Linyang, China's largest electric meter manufacturer, the company received venture capital backing from Citigroup, as well as China's Legend Capital and Hony Capital, prior to its 2006 NASDAQ listing. South Korea's Hanwha Group acquired a controlling stake in 2010. Hanwha had thrived on Europe's solar build-out in the first decade of the 2000s, with sizable sales in Spain and Germany. It took advantage of the distressed European solar market to buy most of Q-Cells, once one of Germany's largest solar manufacturers; the recent focus has been on Japan and China.

HITACHI Smart cities, building management systems, and water treatment plants are among the environmental services that contribute to Hitachi's $100-billion-plus annual revenues. The Japanese conglomerate's 320,000 employees make nuclear power plants and high-speed trains as well as ink-jet printers and washer-dryers. Energy efficiency is part of the company's ethos; environmental projects include a 2013 contract for a $600 million joint venture (with Singapore's Hyflux) seawater desalination plant in Gujarat, India, part of the massive Delhi-Mumbai Industrial Corridor concept backed by Japan and designed to promote energy-saving private-sector-led industrial development in India.

HONDA The world's eighth-largest automobile manufacturer, with fiscal 2014 sales of ¥11.8 trillion ($117 billion), Honda introduced Americans to fuel-efficient cars in the 1970s with its Accord model and remains known for its energy-efficient cars. It has experimented with the use of both hydrogen and electric fuel cells in cars in its efforts to develop sustainable vehicles. In 2013, it announced a co-development agreement with GM for future green technology, which many speculate will center on hydrogen fuel cells. In 2007, Honda began manufacturing solar panels; however, in 2013, it said it would exit the business.

HUTCHISON WHAMPOA Hutchison Whampoa is a Hong Kong–based conglomerate controlled by Li Ka-shing, Asia's wealthiest entrepreneur (*Forbes* puts his 2014 net worth at $34.8 billion). It is a ports-to-telecoms-to-retailer operation with 2013 sales of HK$413 billion ($53.2 billion). In 2008, it set up the Hutchison Water division, which in late 2013 began operations at the world's largest desalination plant, located in Sorek, Israel. The plant (49 percent owned by Hutchison and 51 percent by its local partner) has a capacity is 510,000 cubic meters a day and uses innovative cogeneration technology to cut energy consumption. In 2012, Hutchison Whampoa, which employs 250,000 people in fifty-two countries, bought a water-focused technology incubator in Israel and announced it could invest as much as NIS 100 million ($28.5 million) in start-ups in the following eight years.

HYFLUX Hyflux, a Singapore-listed water services company, was founded and is controlled by Olivia Lum, who immigrated to Singapore from Malaysia in her teens. She subsequently left a research job at Glaxo to found Hydrochem in 1989, initially peddling product by motorcycle in southern Malaysia. Renamed Hyflux, the company today builds and operates water treatment and desalination plants in China, Algeria, and Singapore. Its Magtaa facility in Algeria and its Tuaspring plant in Singapore are among the world's largest saltwater desalination units. The company had 2013 sales of S$536 million ($412 million), and Lum sees future growth opportunities in India and Africa.

HYUNDAI MOTOR GROUP Hyundai, the world's fifth-largest automaker, was spun out of its Korean parent following the 1997–98 Asian financial crisis, a time during which it also acquired struggling Kia Motors. The company reported 87 trillion Korean won ($76 billion) in 2013 sales. It launched its first mass-market hybrid in 2011, and in mid-2014, launched the first mass-market hydrogen-powered car in the U.S. Hyundai's is a version of its Tucson compact, initially sold in southern California, the only place with the hydrogen-station infrastructure to support its engine.

JAIN IRRIGATION Roots of this Indian irrigation company reach back to 1963, when Bharvalal Jain began selling kerosene from a pushcart. Jain Irrigation has become a significant irrigation player; acquisitions have strengthened its capability in water-saving drip irrigation. Jain has received accolades from the World Bank's International Finance Corp. and others for its environmental and social inclusiveness work. The company reported Rs 41.3 billion ($689 million) in sales for fiscal year 2013–14.

Jinko The Li family from China's coastal Zhejiang Province has set up two of China's leading solar panel manufacturers. Brothers Li Xiande and Li Xianhua, along with brother-in-law Chen Kangping, developed separately from another brother, ReneSola (see below) founder Li Xianshou, and formed Jinko. The company began as a panel manufacturer but has now become one of the country's largest solar-farm developers, both inside China and abroad; China Development Bank has promised it $1 billion to finance overseas solar farms from 2013 to 2017. It reported ¥7.08 billion ($1.17 billion) of revenues in 2013.

Keppel Corporation Keppel, today one of Singapore's largest conglomerates, started as a ship repair and port services company in the mid-nineteenth century, when the city was a British colony; now, its businesses include shipbuilding, property, energy, and infrastructure as well as an environmental technology arm. It is a constituent of the Dow Jones Sustainability Index Asia Pacific. The group leads the Singapore side of the Sino-Singapore Tianjin Eco-city 50–50 joint venture. Its sales in 2013 were S$12 billion ($9.7 billion).

Kyocera The Kyocera Group is a Japanese multinational known primarily for its ceramics and electronics parts business, with revenues of ¥1.45 trillion ($14.3 billion) for the year ending March 2014. The group's solar subsidiary brought the 70 MW Kagoshima Nanatsujima Mega Solar Power Plant (Japan's largest) online in November 2013. Japan's generous feed-in tariffs for solar power, designed to fill the gap left by nuclear shutdowns after the Fukushima disaster, have led to a solar boom. Kyocera is increasing solar module production capacity from 800 MW in fiscal 2013 to 1.4GW in fiscal 2015 to take advantage of strong Japanese demand.

Lafarge Lafarge, a Paris-based cement maker, with global capacity twice that of the United States and total revenues in 2013 of €15 billion ($21 billion), is set to become the world's largest cement maker if a proposed merger with Switzerland's Holcim is completed. Its attention is increasingly focused on developing markets in Asia, Africa, and Latin America. The company has pledged to cut its carbon intensity (the amount of carbon emitted per ton of production) by one third from 1990 levels by 2020; it has already cut one quarter. Chairman and CEO Bruno Lafont previously chaired the Cement Sustainability Initiative of the World Business Council on Sustainable Development. The company increasingly uses industrial waste material such as fly ash (from coal-burning power plants) rather than limestone in making concrete.

LDK Founded in 2005, LDK is a vertically integrated company that does everything from producing solar ingots to building solar farms. It reduced its solar cell capacity by 89 percent in 2012, from around 2.2 GW to around 240 MW, but still ranks among China's key solar players. The company has pared its employee headcount from thirty thousand to about ten thousand and has obtained local government support in renegotiating its substantial debt to the China Development Bank and other domestic creditors. In 2014 it defaulted on its offshore debt and filed for bankruptcy. A 22 percent stake in LDK has been acquired by Shunfeng Photovoltaic Technology Co., Ltd., which is positioning itself to benefit from the Chinese government's declaration that it will add 10 GW of solar per year from 2013 to 2015, a substantial boost to demand.

LG The former Lucky-Goldstar, LG is South Korea's fourth-largest *chaebol*, with more than 225,000 employees. Its key manufacturing units are electronics, chemicals, and telecommunications products. LG Chem produces batteries for the Chevy Volt, though its purpose-built Michigan battery plant has had a troubled start because of weak car sales and trouble winning U.S. Environmental Protection Agency approvals. The company has had a green products strategy since the 1990s and has won efficiency awards for TVs and other electronic products.

MAHINDRA & MAHINDRA Mahindra & Mahindra, one of India's largest conglomerates, with $16.5 billion in sales and more than 180,000 employees, has branched out from its core vehicle operations to businesses ranging from IT to aerospace. The company is the world's largest tractor maker, and it acquired electric-car maker REVA with a view to using the company's battery technology in other vehicles. Its solar unit plans more than 150 MW of solar farms for southern India, where it plans to sell power for eight cents a kilowatt hour, which is less than what the state utility charges. Since its efficiency program started in 2007, the company has cut energy and water use. Mahindra Racing is a charter member of the Formula E ('Electric') racing circuit.

MAHINDRAREVA ELECTRIC VEHICLES Electric-car pioneer Reva is based in Bangalore, India. Founder Chetan Maini's fascination with electric cars began when he was an undergraduate at the University of Michigan and raced solar-powered cars across the United States and Australia. Reva is now controlled by Mahindra & Mahindra (see above); in 2013, it introduced a new mass-market model, built at an innovative, energy-efficient factory that incorporates high-tech features in a modestly priced, low-range, low-speed city car. The company has won awards from *The Economist* and *Fast Company* but faces a test in selling its cars in larger numbers.

MANILA WATER In 1997, the Manila Water unit of the Ayala Group, one of the largest and oldest companies in the Philippines, took over the newly privatized water services in the eastern part of the country's capital. Only one out of four families in its service area had round-the-clock access to tap water in 1997; now 99 percent do. Water losses from leakage, theft, and mismanagement have been cut from two thirds to 12 percent, ranking among the world's lowest. The company works with the government on river and watershed clean-ups as part of a national climate-change mitigation strategy; it is expanding to other major cities in the Philippines as well as Vietnam and Myanmar.

MING YANG WIND See *China Ming Yang Wind Power Group*.

MASDAR Masdar is an ambitious desert eco-city developed by Mubadala Development Company, one of oil-rich Abu Dhabi's investment arms. It is a twenty-year, $18-billion-plus project designed by star architect Norman Foster to showcase a variety of leading-edge environmental technologies, ranging from driverless cars to low-energy and low-carbon aluminium and cement. Much of the city's power will be provided by a 10 MW solar plant. A wind funnel towering above Masdar is designed to bring cooler, fresher air into the city. The city owns a 60 percent stake in the 100 MW Shams 1 solar power plant, which opened in 2013.

ORIX Orix, Japan's largest commercial finance company (with a focus on leasing), made its initial environmental services investments in the 1990s. It is now investing ¥114 billion ($1.1 billion) to install 400 MW of solar power, taking advantage of generous Japanese feed-in tariffs introduced in the wake of the 2011 Fukushima nuclear disaster. Other environmental investments include a woodchip biomass plant and wind farms; the company runs an energy conservation business (an energy services company, or ESCO), an electric power trading business, a bulk electricity purchasing business for condominiums, and a storage battery rental service unit, and it is considering geothermal projects.

PANASONIC The Japanese consumer electronics giant Panasonic is also a sizable solar manufacturer; it is installing renewable power, including solar, in new energy-efficient communities in metropolitan Tokyo. It is also building solar farms and rooftop systems in Japan to take advantage of the country's subsidies for solar power. In Bangladesh, Cambodia, Indonesia, Myanmar, and similar markets, the company is introducing consumer products like a solar-powered lantern that doubles as a mobile phone charger. It aims to give away 100,000 solar lanterns by 2018 to some of the 1.3 billion people worldwide who are without electricity.

RENESOLA ReneSola is an integrated polysilicon and solar panel producer started in 2005 by Li Xianshou, a former government official in the Chinese coastal province of Zhejiang. A polysilicon shortage forced it and other Chinese companies to integrate operations, with ReneSola using some of the $1 billion it raised in London, Hong Kong, and elsewhere to build polysilicon furnaces. China Development Bank has helped finance solar farms in Xinjiang and Qinghai as the company broadens from manufacturing to building and operating solar farms. It listed on the New York Stock Exchange in 2008 and reported total revenues of $1.5 billion for 2013. Peak employment of ten thousand had dropped to seventy-nine hundred by end-2013.

SCHNEIDER ELECTRIC Schneider Electric Chairman and CEO Jean-Pascale Tricoire moved to Hong Kong in 2011 to accelerate the group's Asian sales, which in 2013 passed those of Europe to become almost one third of group's total revenues of €24 billion ($33 billion). The company is focused on energy-efficient urbanization; smart grids and energy-efficient buildings are key areas, although its controls and backup systems are ubiquitous. Tricoire notes that with electricity demand doubling in the next twenty years and energy use doubling in the next forty years, coupled with the need to cut global CO_2 emissions in half, the world needs a fourfold efficiency increase.

SHUNFENG PHOTOVOLTAIC INTERNATIONAL LIMITED This Hong Kong-listed company has quietly emerged as an important player in the Chinese solar industry, accumulating controlling stakes in Wuxi Suntech and LDK, struggling solar giants; besides manufacturing, it plans to build solar farms. Zheng Jianming, a researcher-turned-real estate investor, controls Shunfeng; it bought Wuxi Suntech out of bankruptcy—for ¥3 billion ($465 million) cash—and also bought a 21.5 percent stake in LDK, the number two solar wafer manufacturer, in 2012. Shunfeng has funding of at least $3.6 billion from some of China's top banks. As of mid-August 2014 Shunfeng had a stock market valuation of $3.1 billion, ranking it the eighth most valuable among the 128 solar companies tracked by a broad Bloomberg index.

Sembcorp Industries Singapore's Temasek sovereign wealth fund owns half of the Sembcorp Industries conglomerate, one of Singapore's largest. It employs ten thousand people, and the company's businesses range from power plants (mostly coal and gas but also including renewable power in the form of wind, biomass, and waste-to-energy) and water and waste treatment facilities to industrial park development to ship repair. Its energy and water projects are concentrated in China and Southeast Asia but are also found in Chile, South Africa, United Arab Emirates, and Oman. The company posted 2013 sales of S$10.8 billion ($8.6 billion).

Siam Cement Group (SCG) Founded by the Thai monarchy in 1913 to help jump-start the country's modernization, SCG is today one of Southeast Asia's largest and most well-regarded companies. Its 2013 sales totaled $1.36 billion; chemicals make up half of its business, and other units include building materials, paper, and cement. The Dow Jones Sustainability Index named SCG's Building Materials & Fixtures unit the best in its sector in 2011, 2012, and 2013. SCG is a member of the World Business Council on Sustainable Development's Cement Sustainability Initiative.

Singapore Public Utilities Board (PUB) Water is an existential issue for the tiny island-state of Singapore. Long-range policies in Singapore have turned water from a vulnerability into an advantage: a comprehensive Four Taps strategy ensures the country's unhampered access to water. Historically dependent on Malaysia for water, Singapore's water today comes from desalination, recycled water (NEWater), local reservoirs (two thirds of the country is water catchment), and imported water, mostly from Malaysia. The PUB is at the center of a national effort to be a self-styled "hydrohub." Singapore has twenty-five water-related research institutes and laboratories. The government has pledged S$470 million ($375 million) for industry support.

Sino-Singapore Tianjin Eco-city This new city built on the outskirts of Tianjin, one of China's largest and most important cities, is the product of high-level cooperation between China and Singapore and is designed to show what can be done in water-short, degraded wetlands. The city, half the size of Manhattan, is designed to be home to 350,000 people when it is completed in 2020. Hard targets, in the form of twenty-six key performance indicators, are designed to ensure adherence to genuine sustainability at the project, whose first phase is estimated to cost $8 billion.

Sinovel Wind Group Company Limited Beijing-based, Shanghai-listed Sinovel was backed by New Horizon Capital, co-founded by Wen Yunsong, son of former Chinese Premier Wen Jiabao. Once the largest Chinese wind company, it has seen sales slip and profits disappear. In 2013, the U.S. Department of Justice brought an intellectual property theft case against Sinovel, alleging that it bribed an employee of supplier AMSC for access to source code for wind turbine electrical control systems. That year also saw the company admit that it overstated its revenue 10 percent and profits 20 percent in 2011—the year it went public—prompting the resignation of Sinovel's founding chairman, Han Junliang. Successor Wei Wenyuan left two months later. The next year saw new chairman Wang Yuan also resign, shortly after the company reported a ¥3.45 billion ($556 million) loss for 2013.

SoftBank Masayoshi Son (whose 2014 net worth was $19.7 billion, according to *Forbes*), CEO of Japanese telecom and media giant SoftBank (with $5.2 billion in profits on sales of $66 billion in 2013), formed SB Energy in October 2011 to build solar farms and other renewable power sources after the Fukushima nuclear disaster. The intense, high-profile Son has made it his post-Fukushima mission to build a more reliable energy supply for Japan, primarily based on renewables. The target capacity of SB Energy's solar farms is around 260 MW; in a joint venture with Mitsui, the company started construction of a 111 MW solar farm in Hokkaido in late 2013. SB Energy and Mitsui are also building a 48 MW wind farm.

Songdo Near Seoul's international airport is the $40 billion, greenfield smart city of Songdo. It is majority owned by American developer Gale International, with remaining shares held by Korean steelmaker Posco and Morgan Stanley Real Estate. When complete in 2018, the city is set to have 40 percent of its area reserved for green space; will use sensors to track energy, traffic, and waste disposal; and aims to be the first city in the world with all of its major buildings meeting or exceeding LEED's rating requirements. Part of the challenge is filling those buildings; as of late 2013, less than 20 percent of the city's commercial office space was occupied.

Suntech Once Suntech was the world's largest producer of solar panels, but its main operating unit, based in Wuxi, China, went bankrupt in 2013, due to mistaken bets on polysilicon prices, industry overcapacity, falling prices, and heavy debts. Founded in 2001, the company initially received generous support from the Wuxi government. It was the first

Chinese private company to list on the New York Stock Exchange, and its market value peaked at $16 billion. Founder Shi Zhengrong, a solar visionary, in 2008 was named by *Forbes* as one of China's wealthiest people, with a fortune of $2.9 billion. In August 2014, Hong Kong-listed Shunfeng Photovoltaic bought it out of bankruptcy, for ¥3 billion ($493 million), part of a plan that would complete the restructuring of Suntech's $1.75 billion of defaulted debt.

SUZLON ENERGY LIMITED Suzlon Energy was founded by textile entrepreneur Tulsi Tanti in 1995 to ensure reliable power for his textile company. Based in Pune, it is now India's largest and the world's fifth-largest wind turbine manufacturer, with 24 GW of wind capacity installed worldwide. The Indian market experienced a sharp drop in installations when subsidies for setting up wind generation plants were withdrawn in 2012 but hopes are high that the Modi government will reinstate support. In an effort to return to profitability, the Mumbai-listed company is selling assets, eliminating around three thousand jobs, and restructuring its debt.

TAIWAN SEMICONDUCTOR MANUFACTURING COMPANY (TSMC) The Taiwanese chip maker TSMC revolutionized the global electronics industry by enabling fabless semiconductor companies. Its chip-making facilities meant that electronics makers no longer needed their own semiconductor fabrication plants. The company, which had 2013 revenues of NT$597 billion ($19.3 billion), is using its silicon-fabrication expertise to enter solar manufacturing. In 2012, it opened a thin-film solar fabrication plant with 300 MW of annual capacity; plans are to increase that to 1 GW by 2015, part of the company's plan of "driving solar beyond subsidies." It has a significant commitment to water conservation, green buildings, and limits on CO_2 emissions.

TIANJIN ECO-CITY See *Sino-Singapore Tianjin Eco-city*

TOYOTA Toyota is Japan's largest auto company, with sales of ¥22 trillion ($220 billion) in 2012–13. In 1997, it introduced the Prius hybrid, which combined the traditional internal combustion engine with an electric one, dramatically boosting fuel efficiency. The hybrid motor charges its electric half while driving, eliminating the need to recharge with an electric outlet and reassuring consumers who have "range anxiety," the fear of running out of electric power while driving. By 2014, the company had sold more than six million hybrids, mostly Priuses but increasingly other models, including those in the high-end Lexus line. The company plans to begin selling a hydrogen-cell vehicle in mid-2015.

VESTAS Based in Denmark, Vestas is one of the world's oldest and largest wind turbine manufacturers; it began production in 1979 and has over 60 GW of installed capacity worldwide. Half of its sales go to Europe, 40 percent to the Americas, and the balance to Asia. Vestas and Danish rival NEG Micon had about 35 percent of the global market when they merged in 2004. By 2012, the company's global market share had fallen to around 15 percent due to new product development costs, delays introducing new turbines, and competition from China. Under new management the company returned to profitability in 2013 after two years of losses, and has instituted a collaboration with Mitsubishi Heavy in off-shore wind power.

XINJIANG GOLDWIND SCIENCE & TECHNOLOGY Founded in China's far western city of Urumqi in 1998, Chinese wind pioneer Xinjiang Goldwind Science & Technology is the country's largest turbine manufacturer, with 23 percent of the domestic market, and is the second-largest worldwide, with about 10 percent of the global market as of 2013. Listed on the Shenzhen and Hong Kong exchanges, the company has a worldwide installed base of 19 GW of capacity. It bought its German R&D partner, Vensys, in 2008; Vensys's German head doubles as Goldwind's chief technology officer. In May 2010, Goldwind received a $6 billion credit line from the China Development Bank to help its expansion into international markets; it has built wind farms in the United States and Australia.

YINGLI GREEN ENERGY HOLDING COMPANY LIMITED Yingli stepped onto the global stage with a high-profile marketing campaign during the 2014 FIFA World Cup. Headquartered in Baoding, near Beijing, Yingli is a vertically integrated company, from polysilicon manufacturing to ingot casting to solar cell and module production and assembly. The world's largest solar panel manufacturer measured both by production capacity and shipments, it has sold over 10 GW of modules to customers. 2013 sales were ¥ 13.4 billion ($2.1 billion), but it failed to make an operating profit. It is hampered by a heavy debt burden and agreements to buy polysilicon at an above-market price.

NOTES

Introduction

1. Jonathan Kaiman, "China's Toxic Air Pollution Resembles Nuclear Winter, Say Scientists," *The Guardian*. February 25, 2014, accessed May 1, 2014, http://www.theguardian.com/world/2014/feb/25/china-toxic-air-pollution-nuclear-winter-scientists?CMP=fb_gu. The blog is quoted in Wayne Ma, "Beijing Pollution Hits Highs," *Wall Street Journal*, January 14, 2013, accessed April 28, 2014, http://online.wsj.com/news/articles/SB10001424127887324235104578239142337079994.

2. Kerry A. Emanuel, "Downscaling CMIP5 Climate Models Shows Increased Tropical Cyclone Activity over the 21st Century," *Proceedings of the National Academy of Sciences of the United States of America*, approved [for publication] June 10, 2013, accessed May 1, 2014, http://www.pnas.org/content/early/2013/07/05/1301293110.abstract?sid=9fb226fc-6f82–4b7a-8c91–6ce889da1b1a.

3. "ESM Goh: 'The Singapore Child Is Being Suffocated,'" *Straits Times*, June 21, 2013, accessed May 6, 2014, http://www.straitstimes.com/breaking-news/singapore/story/esm-goh-the-singapore-child-being-suffocated-20130621; "PM Lee's Press Conference on Haze, June 20," *Straits Times*, June 21, 2013, accessed March 6, 2014, http://www.singapolitics.sg/news/pm-lees-press-conference-haze-june-20.

4. The World Bank notes that extreme poverty in East Asia fell from 77 percent in 1981 to 14 percent in 2008. "An Update to the World Bank's Estimates of Consumption Poverty in the Developing World," World Bank, February 29, 2012, http://siteresources.worldbank.org/INTPOVCALNET/Resources/Global_Poverty_Update_2012_02–29–12.pdf.

Britain grew 2.04 percent from 1820 to 1870. Angus Maddison, *Monitoring the World Economy* (Paris: Development Center of the Organisation for Economic Co-Operation and Development, 1995), 255. The calculation of China's growth from 1978-2013 used: World Bank. 2014. Data: GDP Growth (Annual %). Accessed August 29, 2014. http://data.worldbank.org/indicator/NY.GDP.MKTP.KD.ZG. In real terms, the economy grew 28.9 times over the 35 year period, measured by constant 2005 U.S. dollars.

5. The World Health Organization in May 2014 said that the world's ten most polluted cities are all in Asia. Six are in India, three in Pakistan and the final one in Iran. For all the publicity Beijing's bad air has received, when measured by average concentrations of the smallest and most dangerous particulates (PM2.5, or particulate matter 2.5 micrometers or less in size, about 1/30th the width of a human hair), New Delhi's is on average almost three times as bad: http://www.newstatesman.com/jonn-elledge/2014/05/most-polluted-cities-earth-are-not-where-you-think.

6. "Ambient Air Pollution Among Top Global Health Risks in 2010," Health Effects Institute, March 31, 2013, accessed May 24, 2014, http://www.healtheffects.org/International/HEI-GBD-MethodsSummary-033113.pdf.

7. Scientists in California observed the highest-ever level of radioactive sulphur in the atmosphere fifteen days after the reactors at Fukushima were cooled by pumping out seawater. Ryan Flinn, "California Researchers Detect Radioactive Release from Fukushima Plant," Bloomberg News, August 16, 2011, accessed August 23, 2014, http://www.bloomberg.com/news/2011-08-15/california-researchers-detect-radioactive-release-from-fukushima-plant.html.

8. As of 2010. *Asian Development Outlook 2013: Asia's Energy Challenge*. Manila: Asian Development Bank, April 2013.

9. There are many different rankings but the largest GHG emitter by far is China, followed by the United States; there is general consensus that Indonesia, India, and Russia are clustered in a group that comprises the next most significant emitters, a grouping that unfortunately has become more pronounced in recent years as these countries' emissions have grown. More recent data shows that Japan and Germany, although they are among the largest and most industrialized of global economies, have fallen quite some distance back from this group, thanks in part to continuing improvements in energy efficiency as well as weak economic growth. When it comes to solving the issue of GHG emissions, it is these big five—China, the U.S., Indonesia, India, and Russia—that matter the most, accounting for half of annual greenhouse gas emissions. When the European Union is added these half-dozen economic units account for almost 60 percent of GHG emissions. See: World Resources Institute. 2014. Climate Analysis Indicators Tool (CAIT 2.0). "Total GHG Emissions Including Land-Use Change and Forestry-2011." Accessed August 31, 2014. http://cait2.wri.org/wri/Country%20GHG%20Emissions?indicator[]=Total%20GHG%20Emissions%20Excluding%20Land-Use%20Change%20and%20Forestry&indicator[]=Total%20GHG%20Emissions%20Including%20Land-Use%20Change%20and%20Forestry&year[]=2011&sortIdx=1&sortDir=desc&chartType=geo.

For a useful alternative, albeit somewhat older, ranking, see Benjamin K. Sovacool, 2014. "Environmental Issues, Climate Changes, and Energy Security in Developing

Asia." June. Page 3. Manila: Asian Development Bank. ADB Economics Working Paper Series No. 399. Accessed August 30, 2014.

10. "ASIA: Top 10 Deadliest Cyclones," IRIN News, September 23, 2010, accessed October 22, 2013. http://www.irinnews.org/report/90556/asia-top-10-deadliest-cyclones.

11. A study in 2004 that took into account the number of "influenza" cases that occurred afterward estimated deaths at closer to twelve thousand, although government reports in the months following gave estimates of four thousand. Michelle L. Bell, Devra L. Davis, and Tony Fletcher, "A Retrospective Assessment of Mortality from the London Smog Episode of 1952: The Role of Influenza and Pollution," Environmental Health Perspectives, 112, no. 1 (January 2004): 6–8 ; "America's Sewage System and the Price of Optimism," *Time*, August 1, 1969.

12. "Present Status and Promotion Measures for Renewable Energy in Japan: Importance of Renewable Energy." Ministry of Economy, Trade and Industry of Japan, http://www.meti.go.jp/english/policy/energy_environment/renewable/ref1001.html (accessed March 1, 2014); Aya Takada and Chisaki Watanabe, "Solar Farmers in Japan to Harvest Electricity with Crops," Bloomberg, May 27, 2014, accessed May 28, 2014, http://www.bloomberg.com/news/2014–05–26/solar-farmers-in-japan-to-harvest-electricity-with-crops.html.

13. As of 2011. "Energy Intensity—Total Primary Energy Consumption per Dollar of GDP (Btu per Year 2005 U.S. Dollars [Market Exchange Rates])," U.S. Energy Information Administration, http://www.eia.gov/cfapps/ipdbproject/iedindex3.cfm?tid=92&pid=46amp&aid=2 (accessed May 26, 2014).

14. China is likely to domestically install at least 10 GW annually of the worldwide production volume of about 38.5 GW of photovoltaic modules. Arnulf Jäger-Waldau, "PV Status Report 2013," European Commission, p. 3, September 2013, accessed March 8, 2014, http://iet.jrc.ec.europa.eu/remea/sites/remea/files/pv_status_report_2013.pdf.

15. "China's 12GW Solar Market Outstripped All Expectations in 2013," Bloomberg New Energy Finance, January 23, 2014, accessed March 6, 2014, http://about.bnef.com/press-releases/chinas-12gw-solar-market-outstripped-all-expectations-in-2013/; "China Out-spends the US for First Time in $15bn Smart Grid Market," Bloomberg New Energy Finance, February 18, 2014, accessed March 6, 2014, http://about.bnef.com/press-releases/china-out-spends-the-us-for-first-time-in-15bn-smart-grid-market/.

16. Nick Robins, Zoe Knight, and Wai-shin Chan, "2013: The Great Disconnect," HSBC Global Research, January 7, 2013, accessed June 6, 2014, https://www.research.hsbc.com/midas/Res/RDV?ao=20&key=NHoQRor6Qo&n=355330.PDF.

17. Joel Kirkland, "China's Ambitious, High-Growth 5-Year Plan Stirs a Climate Debate," *New York Times* (reprint of story from Climate Wire), April 12, 2011, http://www.nytimes.com/cwire/2011/04/12/12climatewire-chinas-ambitious-high-growth-5-year-plan-sti-12439.html?pagewanted=all.

18. This observation has been made about other transformative industries, but I would like to credit Amory Lovins, who made this point when discussing energy-efficient technologies at a seminar in Tallberg, Sweden, in June 2009.

19. Elizabeth Economy, *The River Runs Black: The Environmental Challenges to China's Future*, 2nd ed. (Ithaca, NY: Cornell University Press [A Council on Foreign

Relations Book], 2010). For Economy's more recent analysis, see "China File: How Responsible Are Americans for China's Pollution Problem?" Council on Foreign Relations, February 28, 2014, accessed March 1, 2014, http://www.cfr.org/china/china-file-responsible-americans-chinas-pollution-problem/p32495; and "China Wakes Up to Its Environmental Catastrophe," Bloomberg Businessweek, March 13, 2014, accessed May 24, 2014, http://www.businessweek.com/articles/2014-03-13/china-wakes-up-to-its-environmental-catastrophe#p1.

20. "2012 Greendex Report," National Geographic, p. 78 (Greendex score versus guilt), July 12, 2012, accessed September 22, 2012, http://environment.nationalgeographic.com/environment/greendex/.

Part I. Energy: Sun, Wind, and the End of Coal

1. Fred Pearce, "The Triumph of King Coal: Hardening Our Coal Addiction," *Yale Environment* 360 (October 31, 2011), accessed March 5, 2014, http://e360.yale.edu/feature/the_triumph_of_king_coal_hardening_our_coal_addiction/2458/.

2. "Ambient Air Pollution Among Top Global Health Risks in 2010," Health Effects Institute, March 31, 2013, accessed May 24, 2014, http://www.healtheffects.org/International/HEI-GBD-MethodsSummary-033113.pdf. For an overview, albeit somewhat dated, of coal miners' deaths, see the China Labor Bulletin, "Bone and Blood: The Price of Coal in China," CLB Research Report No. 6, March 2008. Accessed August 28, 2014. http://www.clb.org.hk/en/files/File/bone_and_blood.pdf.

See also China Labor Bulletin. 2014. "Coal Mine Accidents Decrease as Production Stagnates." April 3. Accessed August 28, 2014.http://www.clb.org.hk/en/content/coal-mine-accidents-china-decrease-production-stagnates.

Coal produces 44 percent of global greenhouse gas emissions; it supplies 30 per cent of global energy. The U.S., China, and India are the top producers and the top consumers of coal: Center for Climate and Energy Solutions. "Coal." Accessed August 28, 2014. http://www.c2es.org/energy/source/coal.

3. The issue of when China's coal use peaks is of vital importance, but remains unsettled. The general consensus is that it will be about 2030. See, for example, BP Energy Outlook 2035. 2014. January. p. 69. Accessed August 31, 2014. http://www.bp.com/content/dam/bp/pdf/Energy-economics/Energy-Outlook/Energy_Outlook_2035_booklet.pdf. For a number of Chinese experts views on their country's coal peak, see Ed King, Responding to Climate Change (RTCC). 2014. "Is China Really Committed to Addressing Climate Change?" January 10. Accessed August 28, 2014. http://www.rtcc.org/2014/01/10/is-china-really-committed-to-addressing-climate-change/.

4. Mike Thomas, "The Opportunity for Coal in the Context of Natural Gas" (paper presented at the Coaltrans China Conference, Shanghai, April 10, 2014; Hong Kong: The Lantau Group). The following section on natural gas draws extensively on Thomas's presentation.

5. Hydropower is also a significant part of Asia's, and China's, energy mix, although the lack of untapped water sources and growing public opposition to the population displacement needed for large-scale dam projects will limit future growth.

6. Michael B. McElroy, Xi Lu, Chris P. Nielsen, and Yuxuan Wang, "Potential for Wind-Generated Electricity in China," Science 325, no. 5946 (September 11, 2009): 1378–1380, doi:10.1126/science.1175706, accessed May 28, 2014, http://www.sciencemag.org/content/325/5946/1378.full.html.

1. The Sun Kings

1. The Shi Zhengrong quote comes from an April 30, 2008, speech at the Asia Society in Hong Kong (where he told the restaurant anecdote); my thanks go to Penny Tang at the Asia Society for helping me locate a recording of this speech. Company details were found in company filings, press releases, and regulatory documents; specific references are cited below. I want to thank Jill Baker for her thorough and thoughtful analysis of Suntech's financial statements. This account also draws extensively from Bill Powell and Charlie Zhu, Reuters, "Special Report: The Rise and Fall of China's Sun King," May 18, 2013, accessed October 9, 2013, http://www.reuters.com/article/2013/05/19/us-suntech-shi-specialreport-idUSBRE94I00220130519.

2. For a good overview of the growing global importance of solar power, see David Frankel, Kenneth Ostrowski, and Dickon Pinner, "The Disruptive Potential of Solar Power," *McKinsey Quarterly* 2014, no. 2, 50–55. For China, see PwC. 2012, "2012 Photovoltaic Sustainable Growth Index," especially p. 3, October, accessed March 6, 2014, http://www.pwc.com/en_US/us/technology/assets/pwc-pv-sustainable-growth-index.pdf.

3. Jeffrey Ball, "China's Solar-Panel Boom and Bust," Stanford Graduate School of Business, June 7, 2013, accessed May 16, 2014, https://www.gsb.stanford.edu/news/headlines/chinas-solar-panel-boom-bust.

4. Evidence of favored treatment for Suntech is found in corporate filings with the U.S. Securities and Exchange Commission. Company filings cite "government grants," such as capital grants for the acquisition of equipment, and preferential tax treatment. Suntech China, which was registered in a high-tech zone in Wuxi, qualified as a "high or new technology enterprise." As a result, it was entitled to a preferential enterprise income tax rate of 15 percent so long as it continued to operate in the high-tech zone and maintained its high or new technology enterprise status. It also enjoyed a tax exemption for its first two profitable years of operation, 2003 and 2004, and a preferential tax rate of 7.5 percent for the next three years. Suntech Power Holdings Co., Ltd., FY05 Q3 Form 20-F for the Period Ending December 31, 2005 (filed April 27, 2006), pp. 29, 33, 58, and 60, SEC EDGAR Database, accessed May 26, 2014, https://www.sec.gov/Archives/edgar/data/1342803/000114554906000550/h00483e20vf.htm.

5. "The cost of a solar cell is about 41 U.S. cents a watt today, down from $1.46 in 2010 and about $3 in 2004 when Germany started offering its incentives," according to data from Bloomberg New Energy Finance. "Chinese Zombies Emerging After Years of Solar Subsidies," Bloomberg News, September 9, 2013, accessed March 6, 2014, http://www.bloomberg.com/news/2013-09-08/chinese-zombies-emerging-after-years-of-solar-subsidies.html.

6. Much of the growth in solar power was driven by subsidies, notably in Spain, Germany, and Italy. These three countries each decided to boost solar power by guaranteeing the purchase price at which they would buy solar power that was fed into their electrical grid systems. This boosted the solar industry but also had the unintended effect of jump-starting the Chinese solar industry. Those generous programs fell victim to the global financial crisis. Spain had been among the countries most eager to fund a substantial solar industry. Next came Germany. Germany was not bankrupt, but its program was in some sense too successful—it ended up being very expensive, and its government approved cuts to solar subsidies in March 2012. Tony Czuczka, "Germany Cuts Solar Aid to Curb Prices, Panel Installations," Bloomberg News, March 29, 2012, accessed March 6, 2014, http://www.businessweek.com/news/2012-03-29/germany-cuts-solar-aid-to-curb-prices-panel-installations.

7. Suntech's market capitalization peaked at $13.29 billion on December 26, 2007, when the stock closed at $88.35 (150.4615 million shares outstanding). At its peak, Suntech had soared 589 percent from its $15 IPO price. Price per share and shares outstanding are from Suntech Power Holdings Co., Ltd., FY11 Form 20-F for the Fiscal Year Ended December 31, 2011 (filed April 27, 2012), p. 44, SEC Next-Generation EDGAR System, accessed March 5, 2014.

8. Sales figures are from FY11 Form 20-F for the Fiscal Year Ended December 31, 2011, p. 4, SEC Next-Generation EDGAR System, accessed August 20, 2014, http://www.sec.gov/Archives/edgar/data/1342803/000110465912029568/a12-6915_120f.htm, and Suntech Power Holdings Co., Ltd., FY06 Form 20-F for the Fiscal Year Ended December 31 (filed June 18, 2007), p. 5, SEC Next-Generation EDGAR System, http://www.sec.gov/Archives/edgar/data/1342803/000114554907001069/h01290e20vf.htm.

9. From its December 2005 IPO and a follow-on offering in May 2009, Suntech raised $743 million from the overseas public equity market and another $1.075 billion in the public debt market; it separately sold a $50 million convertible bond to the World Bank–affiliated International Finance Corporation. This is in addition to borrowing from Chinese banks: Suntech financed most of its business with short-term unsecured bank loans.

10. Charles Bai, interviews with the author, May 27, 2013 (in Beijing), and November 4, 2013, and May 8, 2014 (by telephone).

11. Author interview, October 10, 2013, Hong Kong.

12. "Polysilicon Prices Hit Record Low in 2011; Will Head Even Lower Enabling $0.70 per Watt Solar Panels in 2012," BusinessWire, press release, January 19, 2012, accessed March 8, 2014, http://www.businesswire.com/news/home/20120119005221/en/Polysilicon-Prices-Hit-Record-2011-Head-Enabling.

13. "We terminated or significantly amended nearly all of our multi-year supply agreements. For example, in 2011, we terminated our multi-year supply agreement with

MEMC Electronics Materials, Inc. ("MEMC"), for which we recorded an accounting charge of $120 million. However, we still purchase raw materials through supply agreements. If the prices of polysilicon or silicon wafers continue to decrease in the future, we may not be able to adjust our materials costs or manage our cost of revenues effectively. In the event that our raw material costs become higher than that of our competitors who are able to procure polysilicon and silicon wafers at lower prices, our business and results of operations could be materially and adversely affected. In the event we acquire more raw materials than we can fully utilize pursuant to these supply agreements, we may have significant inventory build-up and may have to make further provisions for our commitments and inventory write-downs, which could have a material adverse effect on our business, financial condition, results of operations and prospects. In addition, during the course of renegotiating these supply agreements, we may be subject to litigation if mutual agreement cannot be reached between us and our suppliers." Suntech Power Holdings Co., Ltd., FY11 Form 20-F for the Period Ending December 31, 2011 (filed April 27, 2012), p. 44, SEC Next-Generation EDGAR System, accessed March 5, 2014, http://www.sec.gov/Archives/edgar/data/1342803/000114554907001069/h01290e20vf.htm.

14. Suntech Power Holdings Co., Ltd., 2011 Annual Report, April 30, 2012, p. 5, http://ir.suntech-power.com/phoenix.zhtml?c=192654&p=irol-reportsAnnual.

15. $337.5 million, $142.6 million, and $335.6 million in 2008, 2009, and 2010, respectively. Suntech Power Holdings Co., Ltd., FY10 Form 20-F for the Period Ending December 31, 2010 (filed June 6, 2011), p. 35, SEC Next-Generation EDGAR System, accessed March 5, 2014, http://www.sec.gov/Archives/edgar/data/1342803/000095012311047433/h04443e20vf.htm.

16. In 2006, the first full year after Suntech sold shares on the NYSE, Germany and Spain accounted for 65 percent of its revenue. In 2011, the last year for which the company filed annual financial data, the two countries made up 25 percent of its revenues. The 2006 figures are from Suntech Power Holdings Co., Ltd., FY06 Form 20-F for the Fiscal Year Ended December 31, 2006 (filed June 18, 2007), p. 29, SEC Next-Generation EDGAR System, accessed March 5, 2014, http://www.sec.gov/Archives/edgar/data/1342803/000114554907001069/h01290e20vf.htm. The 2011 figures are from Suntech Power Holdings Co., Ltd., FY11 Form 20-F for the Period Ending December 31, 2011 (filed April 27, 2012), p. 12, SEC Next-Generation EDGAR System, accessed March 5, 2014, http://www.sec.gov/Archives/edgar/data/1342803/000114554907001069/h01290e20vf.htm.

17. Although its financial position looked healthy enough at the time of its initial public offering in 2005, Suntech's financials quickly deteriorated. A large loss in 2007 was followed by low operating margins—in three of the five years in which the company managed to turn a profit, its net income as a percentage of sales ranged between 1.7 percent and 8.2 percent, despite paying almost no taxes. Margin data is from Suntech Power Holdings Co., Ltd., FY11 Form 20-F for the Fiscal Year Ended December 31, 2011 http://www.sec.gov/Archives/edgar/data/1342803/000110465912029568/a12-6915_120f.htm#Item3_KeyInformation_193234, accessed August 18, 2014.

18. On December 26, 2007, market capitalization was $13.29 billion. Market capitalization was approximately $70 million at its low in late March 2013.

19. A Bloomberg report in late 2013 cited $47.5 billion of credit lines that served to cripple the industry with overcapacity. "Solar Defaults Shock Holders as $8.4 Billion Due: China Credit," Bloomberg News, September 3, 2013, accessed May 26, 2014, http://www.bloomberg.com/news/2013-09-03/solar-defaults-shock-holders-as-8-4-billion-due-china-credit.html.

20. Michael Pettis writes extensively on the impact of financial repression in China. See, for example, this blog post: Michael Pettis, "Monetary Policy Under Financial Repression," December 20, 2013, accessed February 4, 2014, http://blog.mpettis.com/2013/12/monetary-policy-under-financial-repression/.

21. Suntech's 2006 Annual Report details some of the tax benefits. Suntech Power Holdings Co., Ltd., FY06 Form 20-F for the Fiscal Year Ended December 31, 2006 (filed June 18, 2007), p. 49, SEC Next-Generation EDGAR System, accessed March 5, 2014, http://www.sec.gov/Archives/edgar/data/1342803/000114554907001069/h01290e20vf.htm.

22. For tax rates, see KPMG International Cooperative, "Global Tax Rates Table," accessed October 8, 2013, http://www.kpmg.com/global/en/services/tax/tax-tools-and-resources/pages/corporate-tax-rates-table.aspx.

23. Yu Ran, "A Wealth of Chinese Billionaires Appears on Global Rich List," *China Daily USA*, February 26, 2014, accessed May 8, 2014, http://usa.chinadaily.com.cn/epaper/2014-02/26/content_17307586.htm. The Hurun report somewhat overstates China's new wealth by including forty-nine Hong Kong billionaires in the China total.

24. Cited in "China's Limit on New Solar Factories Seen Driving M&A," Bloomberg News, September 18, 2013, accessed March 8, 2014, http://www.bloomberg.com/news/2013-09-17/china-to-strictly-limit-building-of-more-photovoltaic-capacity.html. The ENF Company Directory listed 519 solar manufacturing companies in China as of May 8, 2014. ENF Ltd., "Solar Panel Manufacturers in China," accessed May 8, 2014, http://www.enfsolar.com/directory/panel/China.

25. PwC, "2012 Photovoltaic Sustainable Growth Index," p. 2, October 2012, accessed March 6, 2014, http://www.pwc.com/en_US/us/technology/assets/pwc-pv-sustainable-growth-index.pdf. PwC is a member firm of PriceWaterhouseCoopers LLP.

26. Yu says that "the steel industry's profitability was just 0.04 per cent in 2012. Indeed, the profit on two tonnes of steel was just about enough to buy a lollipop." Yu Yongding, "Can China Wean Itself off Its Addiction to Investment?" *South China Morning Post*, October 11, 2013, accessed October 12, 2013, http://www.scmp.com/comment/insight-opinion/article/1328691/can-china-wean-itself-its-addiction-investment.

27. Ehren Goossens and Benjamin Haas. 2014. "Mystery Property Tycoon Makes $533 Million Bet on Solar." April 2. Bloomberg. Accessed August 31, 2014. http://www.bloomberg.com/news/2014-03-31/hong-kong-property-tycoon-makes-533-million-bet-on-solar.html.

28. E-mail with the author, August 29, 2014; this e-mail supplemented conversations with the author from 2010 to 2013.

29. In a December 2012 RenewEconomy interview with First Solar CEO James Hughes, he made the following comments regarding grid parity: "Everyone wants to talk about 'grid parity'—I've banned that phrase from the lexicon of First Solar. Electricity has value only at a point in time and a geographic place. There is no magic

number that describes the true economic cost of electricity. You may have a tariff structure that describes it that way, but that is not the reality, and frankly, sophisticated power markets don't operate like that. So you have to look at time of day, season and location to determine the true cost of power, and there are lots of times of day, seasons and locations where solar is economic today without subsidy. So our focus is to find those places, find those times of day, and find those market structures where we can apply ourselves." "Interview: First Solar CEO James Hughes," RenewEconomy.com, December 13, 2012, accessed March 8, 2014, http://reneweconomy.com.au/2012/interview-first-solar-ceo-james-hughes-72086.

30. Dan Self and Jessie Morris, "Lowering the Cost of Solar PV: Soft Costs with Hard Challenges (Part 1 of 2)," Rocky Mountain Institute, September 25, 2013, accessed October 6, 2013, http://blog.rmi.org/blog_2013_09_25_lowering_the_cost_of_solar_PV_part_one. National Renewable Energy Laboratory, "NREL Releases New Roadmap to Reducing Solar PV 'Soft Costs' by 2020," NREL, press release, September 25, 2013, accessed October 6, 2013, http://www.nrel.gov/news/press/2013/3301.html. International Energy Agency, "Trends 2013 in Photovoltaic Applications," November 29, 2013, accessed March 5, 2014, http://www.iea-pvps.org/index.php?id=3&eID=dam_frontend_push&docID=1733.

31. An International Energy Agency report says that 28.4 GW were installed in 2012. The European Photovoltaic Industry Association believes 31.1 GW were installed. International Energy Agency, "Trends 2013 in Photovoltaic Applications," November 29, 2013, accessed March 5, 2014, http://www.iea-pvps.org/index.php?id=3&eID=dam_frontend_push&docID=1733. Marc Rosa, "Global Solar Capacity Tops 100 Gigawatts on Asian Markets," Bloomberg Sustainability, February 11, 2013, accessed March 8, 2013, http://www.bloomberg.com/news/2013-02-11/global-solar-capacity-tops-100-gigawatts-on-asian-markets.html.

32. As of mid-2014 Bloomberg calculated that China had installed 23 GW of solar power. Bloomberg News. 2014. "China Adds Australia-Sized Capacity in Energy Push." August 7. Accessed August 29, 2014. http://www.bloomberg.com/news/2014-08-07/china-add-australia-sized-solar-capacity-in-energy-push.html.

33. U.S. Energy Information Administration, "Japan," accessed October 21, 2014, http://www.eia.gov/countries/cab.cfm?fips=ja

34. Jeff Kingston, "Japan's Nuclear Village," *Asia-Pacific Journal* 10, 37, no. 1 (September 10, 2012), accessed February 4, 2014, http://www.japanfocus.org/-Jeff-Kingston/3822#sthash.h5P4m1Le.dpuf. See also Laura Araki, "Fukushima One Year Later: An Interview with Daniel P. Aldrich," National Bureau of Asian Research, March 6, 2012, accessed February 4, 2014, http://www.nbr.org/research/activity.aspx?id=219#.UxsBe9zzZyM: "This disaster has radically altered the political landscape in the field of nuclear energy. For the last six decades, the nuclear industry has remained a closed 'iron triangle' or 'nuclear village,' with strong connections between regulators, politicians of the Liberal Democratic Party (LDP), and the private industry." There is a heated debate over nuclear power in Japan even, or perhaps especially, among the country's elite. In September 2013, one of the country's top executives told the author that the government was being unrealistic about trying to cut back on nuclear power; the next day

a prominent retired CEO told the author that building nuclear power plants without proper waste disposal is like building a house without a toilet.

35. The three energy scenarios drawn up by Japan's National Policy Unit can be summarized as follows: the 0% scenario calls for 35 percent renewable energy and 65 percent fossil fuels; the 15 percent scenario calls for 15 percent nuclear energy (a reduction of from the pre-Fukushima 26 percent), 30 percent renewable energy, and 55 percent fossil fuels; and the 20–25 percent scenario envisions only a slight reduction in Japan's reliance on nuclear power for one-quarter of the country's electricity and calls for 25–30 percent renewable energy and 50 percent fossil fuels. "Options for Energy and the Environment: The Energy and Environment Council Decision on June 29, 2012," National Policy Unit, July 2012, accessed March 8, 2014, https://s3-ap-northeast-1.amazonaws.com/sentakushi01/public/pdf/Outline_English.pdf.

36. Gaetan Masson, Marie Latour, Manoel Rekinger, Ioannas-Thomas Theologitis, and Myrto Papoutsi, "Global Market Outlook for Photovoltaics 2013–2017," European Photovoltaic Industry Association, May 2013, accessed March 8, 2014, http://www.epia.org/fileadmin/user_upload/Publications/GMO_2013_-_Final_PDF.pdf. Japan has set a target for solar to make up 10 percent of domestic primary energy demand by 2050.

37. The Orix section is based on numerous conversations with Chairman and CEO Yoshihiko Miyauchi, beginning in 2007, including a formal interview on solar power on September 14, 2013, in Kyoto; and interviews and e-mails with company officials, notably an interview with Atsushi Murakami on September 6, 2013, in Tokyo, as well as reference to the company website. For specific projects, sources include "Construction Starts on a Mega-Solar Project with a Maximum Output of 11.7 MW in Omuta City, Fukuoka Prefecture," ORIX Corporation, press release, September 4, 2013, accessed March 8, 2014, http://www.orix.co.jp/grp/en/news/2013/130904_ORIXE.html; "Construction to Start on Mega-Solar Project at Tokachi Speedway," ORIX Corporation, press release, July 5, 2013, accessed October 5, 2013, http://www.orix.co.jp/grp/en/news/2013/130705_ORIXE.html, accessed October 5, 2014; "ORIX Acquires Robeco," ORIX Corporation, press release, February 19, 2013, accessed October 5, 2013, http://www.orix.co.jp/grp/en/pdf/news/130219_ORIXE1.pdf; "Japan's Largest Condo Building Implementing Bulk Electric Power Purchasing to Use Green Power," ORIX Corporation, press release, May 28, 2013, accessed October 5, 2013, http://www.orix.co.jp/grp/en/news/2012/120528_OepE.html.

38. This is with the 5 percent consumption tax included, as is the ¥37.8 figure below. Thus, the solar producer gets ¥36 per kilowatt hour, down from ¥40.

39. Orix runs a hotel in Beppu City, Oita Prefecture, that is powered with geothermal energy from a 1.9 MW power plant. It has also announced that it will jointly establish a geothermal plant in Gifu Prefecture with Toshiba, looking to commence in 2015 with a capacity of 2 MW. "Toshiba and ORIX to Develop Geothermal Power Generation Business in Nakao, Okuhida Onsen, Gifu Prefecture," ORIX Corporation, press release, November 19, 2013, accessed March 8, 2014, http://www.toshiba.co.jp/about/press/2013_11/pr1901.htm.

40. Bill Dodson, *China Fast Forward: The Technologies, Green Industries and Innovations Driving the Mainland's Future* (Singapore: John Wiley & Sons Singapore, 2012), pp. 137–141 (solar pollution problems). See also "Toward a Just and Sustainable Solar Industry: A Silicon Valley Toxics Coalition Whitepaper," ASVT Coalition, January 14, 2009, accessed May 26, 2014, http://svtc.org/wp-content/uploads/Silicon_Valley_Toxics_Coalition_-_Toward_a_Just_and_Sust.pdf.

41. China and Taiwan together now account for more than 70 percent of worldwide photovoltaic cell production. Arnulf Jaegar-Waldau, "PV Status Report 2013," European Commission, p. 7, September 2013, accessed March 8, 2014, http://iet.jrc.ec.europa.eu/remea/sites/remea/files/pv_status_report_2013.pdf.

2. Blowin' in the Wind

1. Beth Buczynski, "7 Most Impressive Wind Farms (and Turbines) in the World," Global Wind Energy Council, November 6, 2013, accessed November 10, 2013, http://www.care2.com/causes/7-most-impressive-wind-farms-and-turbines-in-the-world.html.

"Wind power now provides 2.5% of global electricity demand—and up to 30% in Denmark, 20% in Portugal and 18% in Spain." International Energy Agency, "Technology Road Map: Wind Energy, 2013 Edition," accessed March 8, 2014, http://www.iea.org/publications/freepublications/publication/Wind_2013_Roadmap.pdf.

According to the Global Wind Energy Council, China had installed 91.4 GW of wind power as of the end of 2013; the United States had 61 GW. The European Wind Energy Association said that Europe had 117.3 GW installed. "Global Wind Statistics 2013," Global Wind Energy Council, May 2, 2014, accessed May 23, 2014, http://www.gwec.net/wp-content/uploads/2014/02/GWEC-PRstats-2013_EN.pdf; "Statistics," European Wind Energy Association, February 2014, accessed May 23, 2014, http://www.ewea.org/statistics/.

2. International Energy Agency, "Technology Roadmap: Wind Energy, 2013 Edition," p. 1.

3. Ryan Wiser and Matt Bolinger, "2012 Annual Wind Market Report," U.S. Department of Energy, p. 1, August 2013, accessed March 8, 2014, http://www.windpoweringamerica.gov/pdfs/2012_annual_wind_market_report.pdf. The Bloomberg New Energy Finance Wind Turbine Price Index had fallen from €1.21 million per megawatt at its height in 2009 to €910,000 per megawatt at the end of 2011 for turbines for delivery in 2013. "Overcapacity and New Players Keep Wind Turbine Prices in the Doldrums," Bloomberg New Energy Finance, press release, March 6, 2012.

4. If a turbine was, as a demonstration, put on a vertical shaft, so that its blades would spin horizontally, it would be unable to rotate freely inside any major league baseball park. As of 2012, the largest commercial wind turbine available is 7.5 MW, with a rotor diameter of 127 meters, and several-larger diameter turbines are available (up to 164 meters). International Energy Agency, "Technology Road Map: Wind Energy, 2013 Edition," p. 13.

5. For its 2013–17 forecasts, refer to Global Wind Energy Council, "Global Wind Report Annual Market Update 2013." 2014. p. 28, April, accessed August 26, 2014, http://www.gwec.net/wp-content/uploads/2014/04/GWEC-Global-Wind-Report_9-April-2014

6. Wiser and Bolinger, "2012 Annual Wind Market Report," p. 20.

7. Michael B. McElroy, Xi Lu, Chris P. Nielsen, and Yuxuan Wang, "Potential for Wind-Generated Electricity in China," *Science* 325, no. 5946 (September 11, 2009): 1378–1380, doi:10.1126/science.1175706, accessed May 28, 2014, http://www.sciencemag.org/content/325/5946/1378.full.html. China's total electricity-generating capacity at the end of 2012 was 1,145 GW, of which 758 GW (66 percent) was from coal-fired power plants, with wind accounting for 61 GW (5.3 percent). Huang Qili, "The Development Strategy for Coal-Fired Power Generation in China," Cornerstone, accessed March 8, 2014, http://cornerstonemag.net/the-development-strategy-for-coal-fired-power-generation-in-china/. In subsequent e-mail correspondence with the author, Michael McElroy made the following points: "The potential [wind] resource could supply 100% of China's future demand for electricity. But, there are serious issues as to how much electricity could be realistically supplied from this resource given its intrinsic variability and the fact that much of the resource is physically removed from major demand centers. A similar situation applies also for the U.S. Analysis in this case suggests that it would be reasonable in the short term to contemplate that wind would supply as much as 20% of demand at acceptable cost by 2030. My view is that a similar situation could apply also for China, assuming that the necessary required investments are made in the power distribution system." E-mail received August 25, 2014.

8. As with so many other fields, China in recent years has gone from a situation where its wind turbine manufacturers are neophytes with limited technological capabilities to one where its manufacturers dominate their own market—the world's largest—and have significant export sales as well. In 2004, Chinese companies and Sino-foreign joint ventures had a modest 25 percent of the domestic wind turbine market. Thanks to a series of preferential policies, that share more than tripled, to 87 percent, in 2009. Joanna Lewis, *Green Innovation in China: China's Wind Power Industry and the Global Transition to a Low-Carbon Economy* (New York: Columbia University Press, 2012), p. 108, figure 4.11.

9. For the cash prize, see ibid., p. 56.

10. For a brief overview of the development of the wind power industry in China, see Liming Qiao, "China's Wind Development: Experiences Gained and Lessons Learnt" (presentation at IRENA Workshop, Copenhagen, April 2012), accessed May 28, 2014, https://www.irena.org/DocumentDownloads/events/CopenhagenApril2012/5_Liming_Qiao.pdf.

11. Lewis, op. cit., p. 108, Figure 4.11 has details on the decline in foreign turbine makers' China market share, from 75 percent in 2004 to 10 percent in 2010.

12. See also the general overview on China from Global Wind Energy Council, "Global Wind 2010 Report: Annual Market Update," p. 31, April 2011, accessed March 8, 2014, http://www.indianwindpower.com/pdf/Global_Wind_2010_Report.pdf.

"Goldwind's newly installed capacity and accumulated installed capacity ranked second in the PRC, and the newly installed capacity ranked fourth internationally."

Goldwind Science & Technology Co., Ltd., 2010 Annual Report, p. 2, April 20, 2010, accessed March 9, 2014, http://www.goldwindglobal.com/web/investor.do.

13. The first wind turbine in China that was connected to the grid appears to have been in Shandong in 1986. This is cited in Joanna Lewis, *Green Innovation in China: China's Wind Power Industry and the Global Transition to a Low-Carbon Economy* (New York: Columbia University Press, 2012), p. xix, p. 82; see also fn. 20. For more on the development of China's wind power industry, see also Shi Lishan, *Three Decades of Wind Power in China* (Beijing: Chinese Renewable Energy Industries Association, 2010).

14. "Through unswerving efforts in developing new and renewable energy sources, China endeavors to increase the shares of non-fossil fuels in primary energy consumption and installed generating capacity to 11.4 percent and 30 percent, respectively, by the end of the 12th Five-Year Plan." "Full Text: China's Energy Policy 2012," *Xinhua*, October 24, 2012, accessed November 23, 2013, http://news.xinhuanet.com/english/china/2012-10/24/c_131927649.htm. This same report notes the following: "The country's installed hydropower generating capacity is expected to reach 290 million kW by 2015... The energy consumption of four major energy-intensive industries—steel, nonferrous metals, chemicals, and building materials—accounts for 40 percent of the national total. Low energy efficiency results in high energy consumption for every unit of GDP." See also, Global Wind Energy Council. 2014. "Global Wind Report: Annual Market Update 2013." P. 17, pp. 42–45.

15. The prediction that foreign turbine makers' market share will fall from 5 percent to 1 percent and other citations from Shen Dechang are from: Bloomberg News. 2014. "China's Wind Turbine Makers Face Market Consolidation." April 18. Accessed August 31, 2014. http://www.bloomberg.com/news/2014-04-17/china-s-wind-turbine-makers-face-consolidation-as-glut-lingers.html.

16. Operating margins and profitability figures are from the proprietary Bloomberg index BRWINDV. Accessed August 27, 2014. The index tracks 25 publicly traded wind-only companies that at the time of writing had a combined valuation of $25.7 billion. The largest constituent company is Vestas; Goldwind is second. Six of the ten companies on the index are Chinese. Thanks to Jill Baker for her research and summary of this index.

17. Global Wind Energy Council, Global Wind Report: Annual Market Update 2013, accessed August 28, 2014, pp. 19, 42. "Country Comparison: Electricity: Installed Generating Capacity," *CIA World Factbook*, accessed August 28, 2014, https://www.cia.gov/library/publications/the-world-factbook/rankorder/2236rank.html.

18. As of 2011, large-scale central-government-administered enterprises and local state-owned enterprises were still the major players in China's wind farm development, with close to 90 percent of all wind power projects invested in, constructed, and completed by these corporations. In 2011, the top five manufacturers in China's newly installed wind power market were Goldwind Science & Technology, Sinovel, United Power, Mingyang, and Dongfang Turbine, respectively. Guodian United Power Technology Company Limited installed 2,847 MW in 2011—a growth of 73 percent over the previous year. Li Junfeng et al., "China Wind Energy Outlook 2012," Global Wind Energy Council, November 2012, accessed March 9, 2014, http://www.gwec.net/wp-content/uploads/2012/11/China-Outlook-2012-EN.pdf.

19. The Goldwind section is based primarily on the following sources: Author interview with Wu Gang, November 19, 2013, Beijing; Lewis, *Green Innovation in China*, pp. 121–144; Xinjiang Goldwind Science and Technology Co., Ltd. IPO prospectus; and "China's Most Powerful People 2009: Wu Gang," *BusinessWeek*, accessed November 15, 2013, http://images.businessweek.com/ss/09/11/1113_business_stars_of_china/28.htm. Also see Xinjiang Goldwind Sci&Tech-A (002202: Shenzhen) "People" (Gang Wu), *Bloomberg Businessweek*, http://investing.businessweek.com/research/stocks/people/person.asp?personId=37616794&ticker=002202:CH, accessed November 15, 2013.

20. Author interview with Wu Gang, November 19, 2013, Beijing. A number of turbines were bought around this time by various Chinese entities. Goldwind deputy director Kathryn Tsibulsky, citing Lishan, wrote the author (e-mail communication, November 26, 2013): "There were a significant number of demonstration projects during the 1970s–1990s. Those installations used both domestic and imported models and they were spread throughout China (Fujian, Zhejiang, Shandong, Xinjiang, Beijing, etc.). Most of those demonstration projects, including many small-scale installations in Xinjiang, were not connected to the grid and generated distributed power. It appears that the first wind farm to be connected to the grid was in Shandong in 1986. That wind farm used 3 Vestas 55kW turbines. Commercial-scale wind farms were developed starting in the mid-2000s."

21. Much of the discussion is based on Lewis's chapter on Goldwind in *Green Innovation in China*, pp. 121–144. Additional detail was provided by Goldwind deputy director Kathryn Tsibulsky, in an e-mail communication to the author, November 26, 2013.

22. "Recharging China's Clean Energy Dream," *Xinhua*, October 2, 2010, accessed March 8, 2014, http://news.xinhuanet.com/english2010/indepth/2010-10/02/c_13539987.htm. See also Lewis, *Green Innovation in China*, p. 123.

23. Goldwind 2013 Annual Report, no page [p. i] and p. 10. See also "Country Comparison: Electricity: Installed Generating Capacity." *CIA World Factbook*, accessed August 23, 2014, https://www.cia.gov/library/publications/the-world-factbook/rankorder/2236rank.html. By way of comparison, California's wind, solar, and geothermal resources make up 15,000 MW of the state's generation mix. "California Electric Grid Sets Solar Generation Record," Reuters, March 10, 2014, accessed May 26, 2014, http://www.reuters.com/article/2014/03/10/utilities-california-solar-idUSL2N0M724F20140310.

24. In addition to China Three Gorges, government shares included Xinjiang Wind Power as well as the Wind Power Research Center and the Solar Energy Co., both controlled by the province's finance department. China Three Gorges (New Energy) is 100 percent owned by SASAC and in turn controls Xinjiang Wind Power. The board of directors is dominated by government and former government officials.

25. The ball bearing issue is mentioned in a 2009 interview with Wu Gang. Lewis, *Green Innovation in China*, p. 130.

26. In 2010, the company struck an agreement with Infineon Technologies AG to introduce core module technology to the company and to produce it in-house. Goldwind also began independent blade design and R&D and in 2010 acquired blade manufacturers Xiexin Wind Power and Xiexin Wind Power, Ltd., to incorporate the blade. Goldwind 2010 Annual Report, p. 11.

27. Michael McElroy, personal e-mail communication, op. cit. "The average capacity factor (the fraction of nameplate potential that is actually realized) for wind installations in the U.S. is 32% as compared to 21% in China. Our analysis suggests that turbine quality can account of about 40% of this difference. A factor unique to China is that for much [for] the time when wind conditions are most favorable (in winter), turbines in China are actually idled sinceexisting coal fired combined heat and power systems must continue to operate to supply hot water required for legislated district heating. The result is reduced demand for the power that could be supplied by wind."

Part II. Our Human World: Cities, Buildings, Wheels

1. Jonathan Woetzel et al., "Preparing for China's Urban Billion," McKinsey & Co., February 2009, accessed May 28, 2014, http://www.mckinsey.com/insights/urbanization/preparing_for_urban_billion_in_china.

2. All Lafont quotes are from an interview with the author, May 30, 2013 (by telephone).

3. See, for example, Lee Kuan Yew's speech and remarks in 2008 where he lays out his vision on what energy and environmental challenges mean for Singapore: Lee Kuan Yew, "Minister Mentor Lee Kuan Yew's Dialogue at Singapore Energy Conference," Singapore Government Press Center, Singapore Energy Conference, Raffles City Conference Center, November 4, 2008, accessed August 13, 2013, http://www.news.gov.sg/public/sgpc/en/media_releases/agencies/mica/transcript/T-20081105-1.html.

3. Cities in a Garden

1. Lee Kuan Yew, *From Third World to First: The Singapore Story: 1965–2000* (Singapore: Singapore Press, Marshall Cavendish Editions, 2006), pp. 200–201 and, more generally, pp. 199–211. The quotation is found on p. 199. For more on the concept of the city in a garden, see "Report: From Garden City to City in a Garden," Center for Livable Cities, Singapore Lecture Series, July 15, 2013, http://www.clc.gov.sg/documents/Lectures/2013/Greening_Singapore_report.pdf.

A concise treatment of the concept is found in "From Garden City to City in a Garden:" http://www.mnd.gov.sg/MNDAPPImages/About%20Us/From%20Garden%20City%20to%20City%20in%20a%20Garden.pdf. Accessed August 4, 2013. See also "Our City in a Garden," Singapore National Parks Board, accessed August 4, 2013, http://www.nparks.gov.sg/ciag/.

2. Robin Lane Fox, "Worth the Gamble," *Financial Times*, March 29, 2013, accessed August 4, 2013, http://www.ft.com/intl/cms/s/2/98778d74-914b-11e2-b4c9-00144feabdc0.html#axzz2azQPpsfp.

3. Ng Lang, *A City in a Garden*, World Cities Summit Issue, Civil Service College, Singapore, 2008, accessed August 4, 2013, http://www.cscollege.gov.sg/Knowledge/Ethos/World%20Cities%20Summit/Pages/08A%20City%20in%20a%20Garden.aspx.

4. Ken Belson, "Importing a Decongestant for Midtown Streets," *New York Times*, March 16, 2008, accessed August 4, 2013, http://www.nytimes.com/2008/03/16/automobiles/16CONGEST.html?ex=1363320000&en=66db1c235736c7ea&ei=5088&partner=rssnyt&emc=rss.

5. Lee Kuan Yew, *The Singapore Story: Memoirs of Lee Kuan Yew* (Singapore: Straits Times Press, 1998), 23.

6. Beijing (17 million), Hong Kong (7 million), Jakarta (26 million), Singapore (5 million), Seoul (22 million), and Manila (22 million) are metropolitan areas that together hold more than 100 million people. Metropolitan Tokyo adds another 37 million people. Add a few other metropolitan areas in Japan (Nagoya, 10 million, and Osaka, 17 million) and China (Shanghai, 21 million; Guangzhou, 17 million; and Shenzhen, 12 million), and the total tops 200 million. Nothing like this scale has ever been seen in human history. It was only in 1968 that the U.S. population passed 200 million. Wendell Cox, "World Urban Areas Population and Density: A 2012 Update," May 3, 2012, accessed July 22, 2013, http://www.newgeography.com/content/002808-world-urban-areas-population-and-density-a-2012-update.

7. Richard Dobbs et al., Urban World: Mapping the Economic Power of Cities (McKinsey Global Institute, March 2011), 14.

8. See more in Jonathan Woetzel, "China's Cities in the Sky," McKinsey & Co., accessed February 9, 2014, http://voices.mckinseyonsociety.com/chinas-cities-in-the-sky/#sthash.ZKqeV2G7.dpuf. This information was supplemented with personal communications (e-mails) with Jonathan Woetzel, February 11–15, 2014.

9. Woetzel et al., *Preparing for China's Urban Billion*, p. 175; personal communications (e-mails) with Jonathan Woetzel, February 11–15, 2014.

10. Dobbs et al., *Urban World*, 12.

11. This skyscraper frenzy is controversial. In the *South China Morning Post*, Stephen Chan reported on a commentary in the official *People's Daily*, as controversy swirled over plans for the Broad Group to build an 838-meter-high Sky City in Changsha. Chan notes that Shanghai Tower is 632 meters tall; it had topped out the previous week. Stephen Chan, "Stop Tall Building Says Party Paper," *South China Morning Post*, August 13, 2012, A6. As of mid-2014, the Changsha Sky City project was listed in the "vision" category in the list of skyscrapers compiled by the Council on Tall Buildings and Urban Habitat. Accessed August 20, 2014. http://skyscrapercenter.com/search.php?var=Sky+city.

12. Woetzel et al., "Preparing for China's Urban Billion," downloaded February 15, 2014. See also Woetzel, "China's Cities in the Sky." Information was also obtained from personal communications (e-mails) with Jonathan Woetzel, February 11–15, 2014.

13. Lee, *From Third World to First*, 199 and, more generally, 199–211 ("Greening Singapore" chapter).

14. Economist Edward Glaeser makes this point eloquently in *Triumph of the City: How Our Greatest Invention Makes Us Richer, Smarter, Greener, Healthier, and Happier* (London: Macmillan, 2011).

15. McKinsey's analysis of cities includes research on China's urbanization. Its early findings were initially published in 2008; full findings were published in March 2009 in

Woetzel et al., "Preparing for China's Urban Billion." In April 2010, McKinsey launched a second report: Sankhe, Shirish, et al. 2010. "India's Urban Awakening: Building Inclusive Cities, Sustaining Economic Growth." McKinsey Global Institute. April. Accessed August 31, 2014. http://www.mckinsey.com/insights/urbanization/urban_awakening_in_india. Also see Dobbs et al., *Urban World.*

The global urban population is growing by sixty-five million annually, equivalent to adding seven new cities the size of Chicago a year. "World Urbanization Prospects, the 2011 Revision," United Nations, Department of Economic and Social Affairs, updated October 7, 2013, accessed March 8, 2013, http://esa.un.org/unup/; Sankhe et al., "India's Urban Awakening," 11.

16. These ideas are from Glaeser, *Triumph of the City.*

17. "Masdar City: Frequently Asked Questions," accessed August 10, 2013, http://masdarcity.ae/en/110/frequently-asked-questions. Masdar is extremely ambitious, and lavishly funded, but it is a small, stand-alone project, more of a laboratory than a city. For an outsider's view, see Tafline Laylin, 2013. "Inhabitat Tours Abu Dhabi's Masdar City." July 7. Accessed September 1, 2014. http://inhabitat.com/exclusive-new-photos-plus-energy-masdar-city-in-abu-dhabi/masdar-city-13/?extend=1.

18. "Nation Building by Design: SOM in China," Skidmore, Owings & Merrill LLP, accessed August 4, 2013, http://www.som.com/ideas/publications/nation_building_by_design_som_in_china; pages 60–71 describe some of SOM's green building work in China.

19. This section is based on interviews with Lim Chee Onn and Ko Kheng Hwa in Singapore in June 2010 and a visit to the Tianjin Eco-city in September 2010; information was also obtained from the site's official website, http://www.tianjinecocity.gov.sg/. At the time of the interview, Ko was CEO of Singbridge. In 2013, he joined Ying Li International Real Estate as group CEO.

Other sources included the following: Sue-lin Wong and Clare Pennington, "Steep Challenges for a Chinese Eco-City," *New York Times*, February 13, 2013, accessed March 9, 2013, http://green.blogs.nytimes.com/2013/02/13/steep-challenges-for-a-chinese-eco-city/; Saleem Ali, "The Promise of Chinese Eco-cities," National Geographic NewsWatch, November 23, 2012, accessed March 9, 2013, http://newswatch.nationalgeographic.com/2012/11/23/eco-cities/; "Horizons: Smart Cities," *BBC News*, accessed March 9, 2014, http://www.bbc.com/specialfeatures/horizonsbusiness/; Coco Liu, "China's City of the Future Rises on a Wasteland," ClimateWire via *New York Times*, September 28, 2011, accessed March 10, 2014, http://www.nytimes.com/cwire/2011/09/28/28climatewire-chinas-city-of-the-future-rises-on-a-wastela-76934.html?pagewanted=all; Julian Wong, "Creating a Better Life: A Closer Look at the Sino-Singapore Tianjin Eco-city Project," Green Leap Forward, November 16, 2008, accessed March 10, 2014, http://greenleapforward.com/2008/11/16/creating-a-better-life-a-closer-look-at-the-sino-singapore-tianjin-eco-city-project/; Jonathan Watts, "China Teams Up with Singapore to Build Huge Eco-city," *The Guardian*, June 4, 2009, accessed March 10, 2014, http://www.guardian.co.uk/world/2009/jun/04/china-singapore-tianjin-eco-city; "Chinese Eco-city Project Gets Boost from Global Environment Facility," World Bank, press release, July 22, 2010, accessed March 10, 2014, http://www.worldbank.org/en/news

/press-release/2010/07/22/chinese-eco-city-project-gets-boost-global-environment-facility; Yuka Yoneda, "Tianjin Eco City Is a Futuristic Green Landscape for 350,000 Residents," Inhabitat, January 10, 2011, accessed March 12, 2014, http://inhabitat.com/tianjin-eco-city-is-a-futuristic-green-landscape-for-350000-residents/.

20. The key performance indicators refer to 2013 and 2020, the target years, respectively, for the start-up phase and completion of the city.

21. Cecilia Tortajada, Yugal Joshi and Asit K. Biswas, *The Singapore Water Story: Sustainable Development in an Urban City-State* (Milton Park: Routledge, 2003), 232.

22. Lee Kuan Yew was intimately knowledgeable about the details of the Suzhou project. At a dinner in Hong Kong on June 6, 1999, we spent an extended period of time discussing the project and his unhappiness with how agreements made by China's leaders were not fully respected by local officials in Suzhou. For the public Q&A session of this event (this does not include remarks about Suzhou), see http://www.businessweek.com/stories/1999–06–20/singapores-lee-you-shape-up-or-perish-extended (accessed August 4, 2013).

23. Lim Chee Onn explained the thinking behind the Jilin project in more detail: "Overlaying it is food security. We want our sources covered well. Historically our policy was simply, don't worry about oil or food. Just worry about money. Oil changed. We made ourselves an oil hub. [In late 2008,] Vietnam and the Thais refused to sell rice. That policy [of ours] now doesn't seem sound. We needed to think about food security."

24. This section is based on a visit to Songdo on March 3, 2014, and an interview with Scott Summers; in addition to the Songdo website, other sources included the following: Sunshine Flint, "Living in: Cities of the future," BBC Travel. June 19, 2013, accessed August 10, 2013, http://www.songdo.com/Uploads/FileManager/BBC%20Travel%206.19.2013.pdf; "New-topias," *Popular Science*, June 2013, p. 66, http://www.songdo.com/Uploads/FileManager/Popular%20Science%206%202013.pdf.

Songdo describes the LEED-certified spaces in the city in a press release from June 27, 2012. "Korea's Songdo International Business District—One of Asia's Largest Green Developments—Surpasses Milestone of 13 Million Feet of LEED Certified Space," Songdo International Business District, press release, June 27, 2012, accessed March 13, 2014, http://www.songdo.com/songdo-international-business-district/news/press-releases.aspx/.

25. Lafarge information is based on numerous company contacts, including an interview with Chairman and CEO Bruno Lafont on May 30, 2013, and a visit to the Andheri East facility on May 3, 2013, which included interviews with Mumbai Construction Development Lab head Lionel Bourbon and Lafarge's country head of marketing Koul Nilesh. Additional information was provided by various Lafarge printed and online materials, including Lafarge India Construction Development Lab, *Bring Innovation Closer to Markets Through Construction System Developments* (Mumbai: Lafarge India, 2013). Lafarge sustainability materials included http://www.lafarge.com/06152012-publication_sustainable-development-2020ambitions_figures_objectives-uk.pdf and http://www.lafarge.com/06152012-publication_sustainable_development-2020ambitions_external_brochure-uk.pdf.

26. Reuters notes the "Cities" code name. Natalie Huet and Caroline Copley, "Holcim, Lafarge Agree to Merger to Create Cement Giant," Reuters, April 8, 2014, accessed May 28, 2014, http://in.reuters.com/article/2014/04/07/lafarge-holcim-idINDEEA3607P20140407.

27. Marie Lechtenberg, "Top 20 Global Cement Companies," *Global Cement Magazine*, December 17, 2012, accessed March 12, 2014, http://www.globalcement.com/magazine/articles/741-top-20-global-cement-companies.

28. About 50 percent of emissions are process emissions occurring in the course of clinker production, some 40 percent arise from the fuel used to heat the cement, and another 10 percent are from electricity use and transportation.

29. "Cement Sector," Invest India, accessed August 14, 2013, http://www.investindia.gov.in/?q=cement-sector.

30. "New York City in 2008 had a total of 3,328,395 housing units, the largest housing stock since the first HVS was conducted in 1965." This excludes vacant units not available for sale or rent. "2010 Housing Supply Report," New York City Rent Guidelines Board, pp. 3–4, accessed August 14, 2013, http://www.housingnyc.com/downloads/research/pdf_reports/10HSR.pdf.

31. Sales in 2013 were 434.25 billion Thai baht ($13.4 billion). "Siam Cement PCL," *Financial Times* market data, accessed April 4, 2014, http://markets.ft.com/research/Markets/Tearsheets/Financials?s=SCC:SET&subview=IncomeStatement.

32. Nonhazardous industrial waste has been eliminated from landfills; 0.01 percent of hazardous waste is still being disposed of in landfills. The 1.85 tons of hazardous waste that were disposed of in landfills in 2013 consisted of insulating material; SCG is researching how to use this as raw material in cement production. SCG, *2013 Sustainability Report*, 45.

33. IPE and the other environmental organizations noted in their report that the cement industry's dust emissions account for about 30 percent of China's total industrial dust emissions and its nitrous oxide pollution accounts for 10 to 12 percent of the country's industrial total. Some of the corporate responses are telling: "China National Building Materials Group Corporation, in holding of CUCC and South Cement, stated that, 'If you [NGOs] have not received a reply to the letter it is probably because the company felt the contents of the letter was of no interest.' Jilin Yatai Group claimed that, 'It was not clear how a reply should be given.' BBMG Corporation replied saying, 'If we feel it's necessary we will follow up and contact you.' To date the only company that has responded showing a willingness to follow up has been Lafarge SA. Other listed companies such as Tangshan Jidong Cement Co., Ltd., Anhui Conch Cement Company Limited and Huaxin Cement Co., Ltd. did not respond at all to inquiries about their records for emissions discharge in breach of regulatory standards."

The Institute of Public and Environmental Affairs, Lvse Jiangnan Public Environmental Concerned Center, Green Hunan, and SynTao. 2013. "Responsible Investment in the Cement Industry: Still a Long Way to Go." June 18. See, especially, pp. 4–6 and 55–62. Accessed August 31, 2014. http://www.ipe.org.cn/Upload/IPE-Reports/Report-Cement-Phase-I-EN.pdf.

34. Author interview on August 19, 2013, supplemented with e-mails on February 13, 2014.

35. Meeting with Mayor Daisaku Kadokawa, July 27, 2013, Kyoto.

4. Buildings for a Greener Asia

1. "Mitsubishi Electric to Install World's Fastest Elevators in Shanghai Tower," Mitsubishi Electric Corporation, press release, September 28, 2011, accessed March 9, 2014, http://www.mitsubishielectric.com/news/2011/0928.html.

2. I have relied extensively on The Global Tall Building Database of the Council of Tall Buildings and Urban Habitat. Note that the rankings of tall buildings change, as some buildings are finished and others are abandoned. I have included three buildings under construction in Russia in the Asian category. Accessed August 29, 2014. http://skyscrapercenter.com/.

3. "Hong Kong's Climate Change Strategy and Action Agenda," Environment Bureau, Hong Kong SAR, p. 15, September 2010, accessed March 9, 2014, http://www.susdev.gov.hk/html/en/council/Paper07–10Annexe.pdf.

4. For an overview of building energy consumption, see United Nations Environment Program, Sustainable Buildings & Climate Initiative. 2009. "Buildings and Climate Change: Summary for Decision-Makers, especially pp. 9–11. Accessed August 29, 2014. http://www.unep.org/sbci/pdfs/sbci-bccsummary.pdf

The International Energy Agency says that buildings account for about 32 percent of final energy or 40 percent of primary energy consumption (final energy measures consumption after losses from, for example, transmission and production). International Energy Agency. FAQs: Energy Efficiency. Accessed August 29, 2014. http://www.iea.org/aboutus/faqs/energyefficiency/.

The American Institute of Architects says that buildings are responsible for as much as 48 percent of greenhouse gas emissions and 68 percent of electricity consumption. American Institute of Architects. Toolkit 2030. Accessed August 29, 2014. https://info.aia.org/toolkit2030/advocacy/architects-green-building.html.

5. Kevin Mo, "From Grey to Green: Make China's Rapid Urbanization Sustainable" (presentation at Natural Resources Defense Council side event during COP 15, Copenhagen, December 8, 2009), 3, 5. Mo says that it would be one month a year, based on a savings of 160 billion kilowatt hours (kWh); the U.S. Energy Information Administration says that annual lighting consumption in the United States is about 461 billon kWh. "Frequently Asked Questions: How Much Electricity is Used for Lighting in the United States?" Accessed August 29, 2014, http://www.eia.gov/tools/faqs/faq.cfm?id=99&t=3

6. Hong Wen, Madelaine Stellar Chiang, Ruth Shapiro, and Mark Clifford, *Building Energy Efficiency: Why Green Buildings Are Key to Asia's Future* (Hong Kong: Inkstone Books, 2007), 10–11. This cites estimates that it is at least four times more expensive to

build new coal-fired capacity than it is to save the same amount of power by making buildings more energy-efficient. This is the negawatts concept, originally developed by Amory Lovins. Justin Fung et al., "From Gray to Green: How Energy Efficient Buildings Can Help Make China's Rapid Urbanization Sustainable," BCG and NRDC, October 2009, accessed May 30, 2014, http://www.bcg.com.cn/en/files/publications/reports_pdf/BCG_From_Gray_to_Green_Oct_2009_tcm42–32257x1x-website.pdf.

7. Barbara Finamore mentions transparency (benchmarking, disclosure, audits); standards (retrocommissioning, lighting upgrades); leading by example (city-owned buildings, public and institutional buildings); and financing/tenant engagement/teaching professionals (maximize leases to help owners recoup upgrade costs from tenants). See Barbara Finamore, "A Tale of Two Cities: Energy Efficient Buildings in New York and Hong Kong," 2013. *Switchboard*, Natural Resources Defense Council, May 14. Accessed June 6, 2014. http://switchboard.nrdc.org/blogs/bfinamore/a_tale_of_two_cities_energy_ef.html.

8. Fung et al., "From Gray to Green," 6, quotes a cost premium in China of 4 to 5 percent; it notes that the Ministry of Science and Technology's 2002 energy-efficient demonstration office building, using more sophisticated technology than most commercial developers would adopt, had a cost premium of 8 percent.

9. Wen et al., *Building Energy Efficiency*, 41–43.

10. For an investor-based approach to green buildings, focused on the European Union, see Institutional Investors Group on Climate Change. 2013. "Protecting Investment Value in Real Estate: Managing Investment Risks from Climate Change." Accessed August 31, 2014. http://www.iigcc.org/publications/publication/protecting-value-in-real-estate-managing-investment-risks-from-climate-chan.

Also see Fung et al., "From Gray to Green," 6–7.

11. "Korea's Songdo International Business District," Songdo International Business District, press release, June 27, 2012, accessed August 15, 2013, http://www.songdo.com/songdo-international-business-district/news/press-releases.aspx/d=386/title=Koreas_Songdo_International_Business_District__One_of_Asias_Largest_Green_Developments__Surpasses_Milestone_of_13_Million_Square_Feet_of_LEED_Certified_Space. See also http://www.songdo.com/Uploads/FileManager/Gale/Press%20Releases/2008/2008.11_%20greenbuild.pdf (accessed August 15, 2013).

12. Kaid Benfield, "As Good and Important as It Is, LEED Can Be So Embarrassing," *Switchboard*, Natural Resources Defense Council, January 18, 2013, accessed December 8, 2013, http://switchboard.nrdc.org/blogs/kbenfield/as_good_and_important_as_it_is.html.

13. Full Esquel sourcing information is available in the discussion of the company in chapter 6.

14. 2nd Green Masterplan, Singapore Building and Construction Authority, http://www.bca.gov.sg/greenMark/others/gbmp2.pdf (accessed May 29, 2014).

15. "BCA's 2nd Green Building Masterplan—80% of Buildings Will Be Green by 2030," Singapore Building and Construction Authority, press release, http://www.bca.gov.sg/newsroom/others/pr270409.pdf (accessed February 7, 2014).

16. Government of Singapore, Building and Construction Authority. [no date] "Singapore: Leading the Way for Green Buildings in the Tropics." pp. 14–15. Accessed August 29, 2014. https://www.bca.gov.sg/greenmark/others/sg_green_buildings_tropics.pdf.

17. "Alliance to Save Energy, October 3, 2013, accessed May 29, 2014, http://www.ase.org/events/2013-evening-stars-energy-efficiency-awards-dinner. See also Government of Singapore, Building and Construction Authority. [no date] "Singapore: Leading the Way for Green Buildings in the Tropics." p. 36. Accessed August 29, 2014. https://www.bca.gov.sg/greenmark/others/sg_green_buildings_tropics.pdf.

18. Sustainability Report 2012, Hang Lung Properties, p. 30, http://www.hanglung.com/Libraries/Document_-_Sustainability/Sustainability_Report_2012.sflb.ashx (accessed May 29, 2014).

19. Ibid., p. 33.

20. Swire Pacific 2013 Annual Report. 2014. Swire Pacific: Hong Kong, pp. 2, 4.

21. Swire Pacific Annual Report 2012, 1, 86. This section primarily focused on Swire Properties. Cathay's other operations are mentioned elsewhere. Water: "Swire Beverages' water efficiency has improved by 39% between 2004 and 2012. Water consumption was reduced by more than three million cbm [cubic meters] in 2012, saving approximately HK$12 million." Ibid., 60; see ibid., 89 for more on water. Fuel: "Cathay Pacific's carbon efficiency relative to its overall capacity (measured in available tonne kilometres) has improved by 12% between 1998 and 2012. This is primarily due to improvements in fuel efficiency as Cathay Pacific acquired new aircraft." Still, from 2003 to 2012 total fuel consumption roughly doubled to about 40 million barrels. Ibid., 12 (for quote and chart on 2003–2012 fuel use). Also see ibid., 88–89 for more on Cathay Pacific (CX). Fish: In 2012, CX and Dragonair said they would stop carrying shark's fin as cargo unless sustainably harvested. See ibid., 101 for more on seafood.

22. Swire Pacific Annual Report 2012, 86.

23. Ibid., 87. Supplementary information and references to the Global Reporting Initiative Guidelines, as well as the Hong Kong Productivity Council audit and recommendations, are at ibid., 220–223. See ibid., 226–229 for information on energy and water by division.

24. Swire Pacific Annual Report 2012, 27. Lightbulb calculation: There are 8,765.81 hours in a year, so a 100 W lightbulb for a year uses 876.581 kWh.

25. Ibid., 81.

26. Ibid., 21. See ibid., 89 for more on Swire Properties.

27. 2011–12 Infosys Sustainability Report, 32. http://www.infosys.com/sustainability/Documents/infosys-sustainability-report-2011-12.pdf. Accessed August 29, 2014. Also see 2012–13 Infosys Sustainability Report, 15, 38–48, http://www.infosys.com/sustainability/Documents/infosys-sustainability-report-2012–13.pdf (downloaded February 7, 2014).

28. Indian commercial buildings typically have EPIs of 200–400. "United Nations Development Program: India Global Environment Facility Project Document," United Nations Development Program, 2011, accessed May 29, 2014, http://www.undp.org/content/dam/india/docs/energy_efficiency_improvements_in_commercial_buildings_project_document.pdf. Infosys's 2013–14 Sustainability Report notes that overall electricity per employee per month declined from 296.5 kWh in fiscal 2008 to 167.2 kWh in

fiscal 2014. Per capita water consumption per month dropped from 3.28 kiloliters (kl) to 2.15 kl. Greenhouse gas emissions (per capita, metric tons of CO_2 equivalent per year) fell from 2.84 to 1.28. Ibid., 39 and 43–46.

29. Infosys is among the companies working with the Lawrence Berkeley Laboratory as part of an Indo-U.S. Joint Clean Energy Research & Development Center, a project that grew out of a 2009 agreement between Indian Prime Minister Manmohan Singh and U.S. President Barack Obama. Gopalakrishnan notes that this collaboration draws on the company's torrent of energy use data that are generated from its extensive corporate facilities. "We have lots of data," he says, allowing building energy-efficiency efforts to be more scientific. Infosys's part of the Indo-U.S. Joint Clean Energy project, which was set up in 2012, focuses on how to integrate information technology with building controls and the physical plant facilities in order to best manage commercial and high-rise residential buildings. The Indian Green Building Center/Confederation of Indian Industries, Schneider Electric, and Natural Resources Defense Council are among the other organizations involved in the project. The Lawrence Berkeley Lab, one of the United States' oldest and most prestigious energy-related research laboratories, has a similar building energy project with Chinese counterparts. Isabel Hilton, "Greening Infosys," China Dialogue, October 14, 2009, accessed July 10, 2013, http://www.chinadialogue.net/article/show/single/en/3285-Greening-Infosys. See also "Berkeley Lab to Lead a U.S.-India Clean Energy Research Center," Lawrence Berkeley Laboratory, press release, April 13, 2012, accessed April 1, 2014, http://newscenter.lbl.gov/feature-stories/2012/04/13/berkeley-lab-to-lead-a-u-s-india-clean-energy-research-center/.

30. "CII Sohrabji Godrej Green Business Center," Asia Business Council, http://www.asiabusinesscouncil.org/docs/BEE/GBCS/GBCS_CII.pdf (accessed August 16, 2013). Also see Hong et al., *Building Energy Efficiency*, 119.

31. Campus information is based on visits on May 3 and December 9, 2013, and interviews with Anup Matthew, Rumi P. Engineer, and others on the campus as well as separate formal and informal interviews with Jamshyd Godrej beginning in May 2007; this information was supplemented with follow-up interviews and e-mails.

32. Indian Green Building Council Scorecard, http://igbc.in/site/igbc/index.jsp (accessed March 8, 2014). Initial information on the Indian Green Building Council came from H. N. Daruwalla, executive vice president and business head, Godrej Electricals & Electronics, December 11, 2012.

33. For a fuller discussion of this issue, see Nirmal Kishnani, *Greening Asia: Emerging Principles for Sustainable Architecture* (Singapore: BCI Asia, 2012).

34. "Tallest Building in China Tops Out: Shanghai Tower Completes Historical Ascension," Gensler, press release, August 2, 2013, accessed March 13, 2014, http://m.gensler.com/about-us/press-release/shanghai-tower-tops-out. For more on the building's energy-efficiency aspirations, see "Shanghai Tower: Gensler Tops Out China's Tallest Building in 2013," Gensler, 2013, accessed June 2, 2014, http://du.gensler.com/vol5/shanghai-tower/#/building-facts; Nicola Davison, "Is the Shanghai Tower the World's First Eco-friendly Skyscraper?" China Dialogue, October 14, 2013, accessed June 2, 2014, https://www.chinadialogue.net/article/show/single/en/6413-Is-the-Shanghai-Tower-the-world-s-first-eco-friendly-skyscraper.

5. Asia on the Move: Cars and Trains

1. "New PC Registrations or Sales," International Organization of Motor Vehicle Manufacturers, http://www.oica.net/wp-content/uploads//pc-sales-2013.pdf (accessed May 28, 2014). When all vehicles are included, the gap narrows, with 22 million and 15.9 million vehicles sold, respectively. "Asia's Shift to Greener Transport Key to Global Sustainability—ADB Forum," Asian Development Bank News and Events, November 6, 2012, accessed May 28, 2014, http://www.adb.org/news/asias-shift-greener-transport-key-global-sustainability-adb-forum.

2. Nearly all of this increase will be in the developing world. "Investing in Sustainable Transport and Urban Systems," Global Environment Facility, December 2012, http://www.thegef.org/gef/sites/thegef.org/files/publication/gef_transportBrch_nov2012_r3.pdf.

3. The $439 billion figure is the official U.S. figure for 2012 United States Trade Representative. 2014. "The People's Republic of China: U.S.-China Trade Facts." Accessed August 30, 2014. http://www.ustr.gov/countries-regions/china-mongolia-taiwan/peoples-republic-china. U.S. Energy Information Administration. 2014. "China is Now the World's Largest Importer of Petroleum and Other Fuels." March 24. Accessed August 24, 2014. http://www.eia.gov/todayinenergy/detail.cfm?id=15531.

4. Malaysia and Brunei are the only two net energy exporters in the region. See "Southeast Asia Energy Outlook," International Energy Agency, p. 22, September 2013, http://www.iea.org/publications/freepublications/publication/SoutheastAsiaEnergyOutlook_WEO2013SpecialReport.pdf.

5. "IMF Calls for Global Reform of Energy Subsidies: Sees Major Gains for Economic Growth and the Environment," IMF, press release, http://www.imf.org/external/np/sec/pr/2013/pr1393.htm (accessed July 20, 2013).

6. Niniek Karmini, "Indonesia Steps Toward Fuel Hike Despite Protests," Associated Press, June 17, 2013, accessed April 4, 2014, http://bigstory.ap.org/article/indonesia-expected-approve-fuel-hike; "Indonesia Fuel Prices Rocket by 44% Sparking Protests," BBC, June 22, 2013, retrieved July 1, 2013, http://www.bbc.co.uk/news/world-asia-23015511. Fuel subsidies were one of the biggest issues confronting President-elect Joko Widowo as he prepared to take office in October 2014. Randy Fabi and Wilda Asmarini. 2014. "Economists Back Jokowi's Fuel Subsidy Plan Cut for Indonesia."Reuters. May 12. Accessed August 30, 2014. http://www.reuters.com/article/2014/05/12/us-indonesia-election-budget-idUSBREA4B03V20140512.

7. Phyllis Schlafly, "Obama's Giveaway to the Communists," WND, January 14, 2013, retrieved April 4, 2014, http://www.wnd.com/2013/01/obamas-giveaway-to-the-communists/; John O'Dell, "New Investor Refocuses Boston Power on China," Edmunds AutoObserver, September 30, 2011, accessed April 4, 2014, http://www.edmunds.com/autoobserver-archive/2011/09/new-investor-refocuses-boston-power-on-china.html; Boston-Power. 2011. "Boston-Power Announces $125 Million in New Financing." September 20. Accessed August 30, 2014. http://www.boston-power.com/news/press-releases/boston-power-announces-125-million-new-financing. Mark Halper, "Japan, S. Korea Dominate Electric Car and Hybrid Battery Market," Smartplanet, February 15,

2012, accessed May 29, 2014, http://www.smartplanet.com/blog/intelligent-energy/japan-s-korea-dominate-electric-car-and-hybrid-battery-market/; Jeff Cobb, "LG Chem Opens 'Largest' EV Battery Plant," Gm-Volt.com, April 15, 2011, accessed April 4, 2014, http://gm-volt.com/2011/04/15/lg-chem-opens-ochong-battery-plant-expects-major-market-share/; Jin Hyunjoo and Christian Hetzner, "Hyundai Revs Up Fuel-Cell Plan as Battery Technology Disappoints," Reuters, September 25, 2012, accessed April 7, 2014, http://www.reuters.com/article/2012/09/25/us-autoshow-paris-hyundai-fuelcell. For more on Hyundai, see Hyundai Motor America. 2014. "Hyundai Fuel Cell Press Releases." Accessed August 30, 2014. http://www.hyundainews.com/us/en-us/FuelCell/PressReleases.aspx.

8. Qin Xingcai's comments and information about the company come from a corporate presentation made at the Standard Chartered Bank Earth's Resources Forum 2013 in Hong Kong on June 20, 2013. State backing, says Qin, included a ¥2 billion investment by state oil company CNOOC in 2009. Before 2009, the company focused on consumer batteries (with customers such as LG, Dell, Microsoft, Lenovo, Samsung, and Apple), but this portion of the business has been declining as it has focused more on vehicles and so-called dynamic batteries, or large-scale power supplies.

9. "Worldwide Sales of Toyota Hybrids Top 6 Million Units," Toyota, press release, January 14, 2014, accessed May 23, 2014, http://corporatenews.pressroom.toyota.com/releases/worldwide+toyota+hybrid+sales+top+6+million.htm.

10. For details of the support provided to electric vehicles as part of China's 2008 stimulus program, as well as details on the program itself, see "Chinese Macroeconomic Management through the Crisis and Beyond," Australian Treasury, 2011, accessed July 1, 2013, http://www.treasury.gov.au/PublicationsAndMedia/Publications/2011/Chinese-Macroeconomic-Management-Through-the-Crisis-and-Beyond/working-paper-2011-01/Chinas-stimulus-package.

11. ChinaAutoWeb. 2014. "Plug-In EV Sales in China Rose 37.9% to 17,600 in 2013." January 10. http://chinaautoweb.com/2014/01/plug-in-ev-sales-in-china-rose-37-9-to-17600-in-2013/.

Tom Woody, "The Chart That Shows Why China Is Desperate to Switch to Electric Cars," *The Atlantic*, May 19, 2014, accessed May 23, 2014, http://www.theatlantic.com/technology/archive/2014/05/the-chart-that-shows-why-china-is-desperate-to-switch-to-electric-cars/371153/; Christopher Marquis, Hongyu Zhang, and Lixuan Zhou, "China's Quest to Adopt Electric Vehicles," *Stanford Social Innovation Review*, Spring 2013, accessed May 23, 2014, http://www.hbs.edu/faculty/Publication%20Files/Electric%20Vehicles_89176bc1-1aee-4c6e-829f-bd426beaf5d3.pdf.

12. "The China New Energy Vehicles Program: Challenges and Opportunities," World Bank and PRTM Management Consultants, Inc., April 2011, accessed May 30, 2014, http://siteresources.worldbank.org/EXTNEWSCHINESE/Resources/3196537-1202098669693/EV_Report_en.pdf.

13. Chetan Maini information is based on author interviews on April 30, 2013, in Bangalore with Maini and other senior executives; this was supplemented by unpublished company presentations as well as the following articles: Shreyasi Singh, "How I Did It: Chetan Maini," India Inc., March 2010, accessed July 17, 2013, http://www.growthinstitute

.in/emagazine/mar10/resource-center.html; "An Electrifying Thinker: Exclusive Interview with Electric Vehicle Pioneer Chetan Maini," Mahindra Rise, September 12, 2011, accessed July 17, 2013, http://rise.mahindra.com/interview-with-electric-vehicle-pioneer-chetan-maini/; Hannah MacMurray, "Interview with Chetan Maini," GreenCarDesign, December 14, 2011, accessed July 17, 2013, http://www.greencardesign.com/site/interviews/interview-chetan-maini-2011; "Zero-Emission Vehicle Legal and Regulatory Activities and Background," California.gov, June 10, 2013, accessed July 17, 2013, http://www.arb.ca.gov/msprog/zevprog/zevregs/zevregs.htm; Manu P. Toms, "Mahindra Reva Targets 30,000 Electric Cars a Year by 2015–16," *Hindustan Times*, August 22, 2012, accessed July 17, 2013, http://www.hindustantimes.com/Autos/HTAuto-TopStories/Mahindra-Reva-targets-30-000-electric-cars-a-year-by-2015-16/Article1-917825.aspx; "Creating a New Ultra Low Emission Discount," Transport for London, https://consultations.tfl.gov.uk/roads/5503a5b6 (accessed July 17, 2013).

14. Sarika Bansal and Skylar Bergl, "Most Innovative Companies: Mahindra Reva, Tesla," *FastCompany*, 2013, accessed June 22, 2013, http://www.fastcompany.com/most-innovative-companies/2013/mahindra-reva-tesla.

15. Mahindra Reva Electric Vehicles Pvt. Ltd., Asset Light Process—Capital Investments 10x Lower than Traditional Automotive Product Development Cycles (unpublished company document dated 2013), 28.

16. Author interview with Wang Chuanfu, October 26, 2012, Shenzhen; BYD Website (http://www.byd.com), especially the 'News' section for details on sales (http://www.byd.com/news/newslist.html). Accessed August 30, 2014; and the following articles: Justine Lau, "Buffett Buys BYD Stake," *Financial Times*, September 30, 2008, accessed April 7, 2014, http://www.ft.com/intl/cms/s/0/235c9890-8de5-11dd-8089-0000779fd18c.html?siteedition=intl#ixzz2ZI7mmh2v; "Buffett's BYD Profit Up to 200% as First Quarter 2013 Profit Triples," *Forbes*, April 26, 2013, accessed April 7, 2014, http://www.forbes.com/sites/gurufocus/2013/04/26/buffetts-byd-profit-up-to-200-as-first-quarter-2013-profit-triples/; "Chuanfu, Wang," Reuters, http://in.reuters.com/finance/stocks/officerProfile?symbol=RENN.N&officerId=1742100 (accessed April 7, 2014); "Wang Chuanfu," *BusinessWeek*, June 8, 2003, accessed April 7, 2014, http://www.businessweek.com/stories/2003-06-08/wang-chuanfu; "Company Profile," BYD, http://www.byd-auto.net/company/profile.php (accessed April 7, 2014). Wang told my former *BusinessWeek* colleague Bruce Einhorn that "the letters had no special meaning, although now he jokes that they stand for 'bring you dollars.'"

The company also topped BW's 2010 Tech 100 list: "BYD Tops Bloomberg Businessweek's 12th Annual Tech 100 List," BusinessWire, May 20, 2010, retrieved July 20, 2013, http://www.businesswire.com/news/home/20100520006751/en/BYD-Tops-Bloomberg-Businessweek%E2%80%99s-12th-Annual-Tech; "Mercedes Set to Launch New Electric Car in China," WorldCrunch translation of *Die Welt* article, December 7, 2011, accessed April 7, 2014, http://www.worldcrunch.com/business-finance/mercedes-set-to-launch-new-electric-car-in-china-c2s4247/#.VDCsv_mSySo; "Most Innovative Companies 2010: BYD," *FastCompany*, accessed June 22, 2013, http://www.fastcompany.com/mic/2010/profile/byd.

17. Quoted in Ezra F. Vogel, *Deng Xiaoping and the Transformation of China* (Cambridge: Belknap Press, 2011), 129. Much has been written of the Shenzhen Special

Economic Zone; for a good summary, see ibid., 129–130, 135–153. See ibid., 136 for the population figure of twenty thousand. The population figures are not strictly comparable because Shenzhen was carved out of the southern part of what had been Baoan County and encompassed two towns and fifteen communes.

18. "It is counterproductive to promote EVs in regions where electricity is produced from oil, coal, and lignite combustion." T. R. Hawkins, B. Singh, G. Majeau-Bettez, and A. H. Strømman, "Comparative Environmental Life Cycle Assessment of Conventional and Electric Vehicles," *Journal of Industrial Ecology* 17 (2013): 53–64, doi: 10.1111/j.1530-9290.2012.00532.x.

19. Axel Krieger, Philipp Radtke, and Larry Wang, "Recharging China's Electric Vehicle Aspirations," McKinsey & Company, July 2012, accessed July 20, 2013, http://www.mckinsey.com/insights/energy_resources_materials/recharging_chinas_electric-vehicle_aspirations.

20. The subsidy figures are found in the BYD annual reports, 2009–2013, under "Note 7, Government Grants and Subsidies." (A simple average exchange rate was used to calculate the approximate US$ equivalent.)

21. The Toyota section is based largely on interviews on August 21, 2013, in Singapore with Bernard O'Connor, executive vice president, Toyota Motor Asia Pacific, and Akiko Machimoto, general manager, Public Relations Department; subsequent e-mail communications; and articles, notably "Toyota's Hybrid Technology" (company photocopy presentation by Satoshi Ogiso; dated May 22, 2013). See also "Toyota to Launch 'New Era' of High MPG Hybrids, Expand Its Global Hybrid Rollout," Toyota press release and related remarks by Satoshi Ogiso, August 28, 2013, accessed May 29, 2014, http://toyotanews.pressroom.toyota.com/releases/2013+toyota+global+hybrid+rollout.htm; Jamshyd Godrej provided valuable insight into Toyota's thinking during the course of numerous formal and informal conversations; the Godrej quote is from an author interview in Mumbai, May 3, 2014. The *Fortune* quote is from Alex Taylor III, "The Birth of the Prius," *Fortune*, February 24, 2006, accessed August 30, 2014, http://archive.fortune.com/magazines/fortune/fortune_archive/2006/03/06/8370702/index.htm.

22. See "Chart of the Day: MRT Ridership Could Grow 89% to 1.3b in 2030," *Singapore Business Review*, February 15, 2013, http://sbr.com.sg/economy/news/chart-day-mrt-ridership-could-grow-89–13b-in-2030.

23. "The MTA Network, Public Transportation for the New York Region," MTA.info, http://web.mta.info/mta/network.htm (accessed June 19, 2013); "Key Facts," Transport for London, http://www.tfl.gov.uk/corporate/modesoftransport/londonunderground/1608.aspx (accessed June 19, 2013).

24. Wang Hui, "China Plans Five-Year Leap Forward of Railway Development," China Features, http://bg.chineseembassy.org/eng/dtxw/t274660.htm (accessed April 9, 2014); Han Qiao, "China's High-Speed Programme Back on Track," *International Railway Journal*, January 10, 2013, accessed May 18, 2013, http://www.railjournal.com/index.php/high-speed/chinas-high-speed-programme-back-on-track.html.

25. The long-term impact of the crash is still uncertain. Bill Dodson 2012. *China Fast Forward: The Technologies, Green Industries and Innovations Driving the Mainland's Future*. Singapore: John Wiley & Sons Singapore, p. 22 argues that the crash will be a

seminal moment in China's development. "The high-speed train accident and the scores of fatalities resulting from the incident proved to be a tipping point in China's hurried modernization. The contract between citizens, the leadership of the country, and domestic commercial interests on the Internet was coming undone." That statement, written not long after the crash, probably reflected the intensity of the moment more than the long-term significance of the accident. It was nonetheless a significant event in China modernization, perhaps akin to the 1967 cabin fire that killed all three Apollo 1 astronauts in their craft on the launch pad at Cape Kennedy. This was a national trauma for the United States, but lessons were learned, and two-and-a-half years later the first astronauts landed on the moon.

Keith Bradsher cites a debt figure of ¥4 trillion, or $640 billion, for the high-speed project in "China Opens Longest High-Speed Rail Line," *New York Times*, December 26, 2012, retrieved April 11, 2014, http://www.nytimes.com/2012/12/27/business/global/worlds-longest-high-speed-rail-line-opens-in-china.html.

26. For this section I relied on an extensive and wide-ranging interview with Glenn Frommer (who subsequently left the company), April 12, 2013, at the MTR offices in Fotan, Hong Kong, as well as the MTR and other materials cited below. The company regularly tops the Hang Seng Corporate Sustainability Index and received the Sustainability Excellence Award in 2012 from the Chamber of Hong Kong Listed Companies. For a full list of recent awards, see "Awards Table," MTR Sustainability, http://www.mtr.com.hk/eng/sustainability/2012rpt/performance-data-awards.php (accessed April 11, 2014).

27. From the 2011 Annual Report, chairman's letter: "During 2011, a new section of our Design Standards Manual focusing on energy efficient railway design has been implemented. It specifies that all new railway projects will consider carbon assessments and will facilitate carbon emissions reduction throughout the project life cycle. An initial assessment of carbon emissions to predict and track emissions has been completed. Tools are now being developed to compare the embodied carbon in our railway infrastructure with our operations, thus allowing a life cycle assessment of carbon emissions. This is a world's first for a railway company." MTR Corporation. Annual Report 2011, p. 10.

28. MTR Corporation, Annual Report 2013, pp. 2, 14 (ridership numbers). http://www.mtr.com.hk/eng/investrelation/2013frpt_e/EMTRAR2013F.pdf (accessed May 23, 2014). The cross-harbor figure is from Frommer interview, op cit.

29. The figures on length of track are from 2013 Annual Report, p. 6. There are various ways of measuring profitability. The mid-2014 value of the government's 77 percent stake in the publicly traded company was $17.7 billion. Bloomberg. 2014. MTR Corp. Ltd 66:HK. Accessed August 30, 2014. http://www.bloomberg.com/quote/66:HK. The Hong Kong government initially injected HK$32.2 billion ($4.1 billion) into the MTR; Dr. Agachai Sumalee, "Rail + Property + Pedestrian Model (RPP): Case Study of Hong Kong and Implication for Thailand," Department of Civil and Structural Engineering, Hong Kong Polytechnic University, http://www.atransociety.com/2014/pdf/pdfSymposium2010/download/ppt_day_1/(3)%20Dr.%20Agachai%20Sumalee.pdf (accessed May 29, 2014). For new capital projects costing HK$160 billion ($20.5 billion), see "South Island Line (East) and Kwun Tong Line Extension to Commence Works Soon," Hong

Kong government, May 18, 2011, accessed April 8, 2014, http://www.info.gov.hk/gia/general/201105/18/P201105180326.htm. These five simultaneous construction projects have strained MTR resources and caused delays; largely as a result, in mid-2014 CEO Jay Walder, previously the head of the New York City MTA, was forced out of his job. For the Express Link, see MTR. 2014. Express Rail Link. Accessed August 30, 2014. http://www.expressraillink.hk/en/home/.

30. For electricity cuts of 15 percent (from 5.68 kWh for each kilometer traveled by one of its subway carriages to 4.83 kWh per kilometer), see MTR Corporation, 2011 Sustainability Report, http://www.mtr.com.hk/eng/sustainability/2011rpt/sustainabilityreport2011.pdf (accessed May 30, 2014). For LED lights, see MTR Corporation, 2011 Annual Report, 33-34, accessed August 30, 2014. http://mtr.com.hk/eng/investrelation/2011frpt_e/EMTRAR2011F.pdf.

Part III. Nature: Forests, Farms, and Water

1. I was told about the Four Seasons evacuation by two people with direct knowledge of the incident. The account of the floods come from various press accounts. See, for example, Ian MacKinnon. 2007. The Guardian. "Four-Meter Floodwaters Displace 340,000 in Jakarta." February 5, 2007. Accessed August 30, 2014. http://www.theguardian.com/world/2007/feb/05/weather.indonesia.

2. Bruno Philip, "Jakarta Faces Up to a High Flood-Risk Future," *The Guardian*, February 5, 2013, accessed May 28, 2014, http://www.theguardian.com/world/2013/feb/05/jakarta-floods-rising-sea-levels.

6. "Water Is More Important than Oil"

1. This story was widely reported, in both domestic and international media. See, for example, "Pig Pollution Shows Urgent Need to Ensure Water Quality," *Global Times*, Zhang Yi, March 3, 2013, accessed January 26, 2014, http://www.globaltimes.cn/content/767692.shtml#.UuS283lm4y4; "191 More Dead Pigs in Huangpu," *SinaEnglish*, March 3, 2013, accessed January 26, 2014, http://english.sina.com/china/2013/0324/574949.html; "China Pulls Nearly 6,000 Dead Pigs from Shanghai River," *BBC News*, March 13, 2013, accessed January 26, 2014, http://www.bbc.co.uk/news/world-asia-china-21766377.

2. Though he may not have originated the joke, this was posted on Sina Weibo by the former head of Google China, Kai-Fu Lee: "Skip the Cigarettes, Breathe in Beijing's Air," Global Voices, March 13, 2013, accessed January 26, 2014, http://globalvoicesonline.org/2013/03/13/china-beijings-free-cigarettes-and-shanghais-pork-soup/. Separately, Singapore Prime Minister Lee Hsien Loong told the joke during a visit to the United States in early 2013: Matthew Pennington, "Singapore PM Draws Laughs in US Speech," Associated Press, April 3, 2013, accessed January 29, 2014, http://news.yahoo.com/singapore-pm-draws-laughs-us-speech-111914557—politics.html.

3. This section draws extensively on Brahma Chellaney's work. Brahma Chellaney, *Water: Asia's New Battleground* (Washington, DC: Georgetown University Press, 2011), 16. Chellaney cites a 2009 UNESCAP study on an Index of Water Available for Development: United Nations Economic and Social Commission for Asia and the Pacific, *Sustainable Agriculture and Food Security in Asia and the Pacific*, 63 (fig. III-2).

4. This paragraph is based on Chellaney, *Water*, 8ff., 11n14, citing UN Population Division. The current number is from UNESCAP, *Statistical Yearbook for Asia and the Pacific 2013* (United Nations Publications: Bangkok: 2013 (October)). See http://www.unescap.org/stat/data/syb2013/ESCAP-syb2013.pdf. Chellaney (*Water*, 36) says that northern China has 64 percent of China's cultivated land but only 19 percent of its water resources.

"Water is more important than oil." Masdar CEO Dr. Sultan Ahmed Al Jaber made the statement, citing earlier remarks he said were made by Sheikh Mohamed bin Zayed, the crown prince of Abu Dhabi. 2012. "His Highness Sheikh Mohamed bin Zayed Launches International Water Summit." January 18. Accessed August 24, 2014. http://www.worldfutureenergysummit.com/Portal/news/18/1/2012/his-highness-sheikh-mohamed-bin-zayed-launches-international-water-summit.aspx.

Young Asians are particularly worried about water. One indication of this came in 2010, when the Asia Business Council sponsored the Asia's Challenge 2020 essay contest, which asked young Asians to write about the biggest obstacle facing the region over the next decade. More wrote about the environment than any other issue. Many wrote about water. Shreyans Jain wrote chillingly about a crowd in an upper-class Delhi neighborhood crowding around a water truck during a drought: "By the time I could reach the water tanker, hundreds of other colony members had already lined up with buckets and pots and mugs in hands. It was impossible to give water to everyone, and soon some people started bribing the official. This situation did not go down well with others who called it unethical, and soon a fight picked up. . . . I could not help but wonder if this was the situation in a supposedly rich area of India's most pampered city, what would be the state of India's villages?" Mark L. Clifford and Janet Pau. *Through the Eyes of Tiger Cubs: Views of Asia's Next Generation*. (Singapore: John Wiley & Sons, 2012), 74.

5. Chellaney, *Water*, 6.

6. One of the more well-known spills in recent years occurred in the Guangxi Longjiang River in early January 2012. Hechi authorities estimated that forty thousand kilograms of fish were found dead after a spill eighty times the legal limit, including twenty tons of toxic cadmium metals, resulted in a slick more than one hundred kilometers long. In January 2014, Greenpeace spotted a vast toxic spill the size of fifty Olympic swimming pools, which could be seen from satellite images, off the coast of Shishi in southern China. It appeared to have originated from a pipe used by the Haitian Environmental Engineering Co. Ltd., which processes wastewater from nineteen fabric-dyeing factories in the area.

7. This section is based on interviews with company executives and employees, including a visit to a number of facilities on August 31, 2010, and an interview with

Gerardo Ablaza on March 12, 2013, in Manila, as well as company annual reports and other publicly available information. I also drew extensively from *Good Practices in Urban Water Management* (Manila: Asian Development Bank and National University of Singapore, 2012), especially chapter 6, "Manila, Philippines," 103–132.

8. SitRep No. 92 "Effects of Typhoon 'Yolanda' (Haiyan)," 2014. Republic of the Philippines. National Disaster Risk Reduction and Management Council, January 14. Quezon City. Accessed August 24, 2014, http://www.ndrrmc.gov.ph/attachments/article/1125/NDRRMC%20Update%20re%20Sit%20Rep%2092%20Effects%20of%20%20TY%20%20YOLANDA.pdf http://reliefweb.int/sites/reliefweb.int/files/resources/NDRRMC%20Update%20re%20Sit%20Rep%2092%20Effects%20of%20%20TY%20%20YOLANDA.pdf. Haiyan followed Ondoy, a September 2009 storm that saw 450 millimeters (almost 18 inches) of rain fall in 12 hours, the sort of storm that is supposed to occur only once every 180 years. *Good Practices in Urban Water Management*, p. 104.

9. Liza Lucero, a community leader in another barangay in metropolitan Manila, noted that she was paying 65 pesos "every other day for water bought from delivery trucks," or about 975 pesos per month. "With Manila Water, our expense was significantly reduced to only about PhP 135 per month. Our savings can now be spent on other important items such as food." Manila Water, 2011 Sustainability Report, "Clean Water for Healthy Communities," 45.

10. The electricity supply was so unreliable that one friend living in Manila at the time told how her baby daughter would wake, hot and uncomfortable, when the power went out in one of the many rolling, scheduled power outages, and mumble "brownout" before restlessly falling asleep again.

11. *Good Practices in Urban Water Management*, 103–132. See also World Bank, "Private Concessions: The Manila Water Experience," March 2010, accessed May 28, 2014, http://siteresources.worldbank.org/NEWS/Resources/ManilaWaterProject3-31-10.pdf.

12. "Manila Water Shares Best Practices in Singapore International Water Week," Manila Water, press release, http://www.manilawater.com/investor/News%20and%20Updates/Pages/ReadNewsItem.aspx?ItemID=4.

13. The 18 percent annual return figure is from Manila Water. 2012. Minutes of the Annual Meeting of Shareholders. April 16, p. 3. The 2012 figure of 67 percent is from Manila Water. 2013. Minutes of the Annual Meeting of Shareholders. April 15, pp. 4–5. In addition to the awards mentioned, Manila Water was one of nine international companies to get Okovision Sustainability Leadership awards in 2012. http://www.manilawater.com/investor/News%20and%20Updates/Pages/ReadNewsItem.aspx?ItemID=47; Accessed August 24, 2014.

The figure on the number of new households and low-income households served by Manila Water is from Manila Water. 2012. "Manila Water now has 889,448 water connections, serves more than 6.2M." November 23. Accessed August 24, 2014. http://www.manilawater.com/investor/News%20and%20Updates/Pages/ReadNewsItem.aspx?ItemID=51.

14. The company's Tubig Para Sa Barangay program, which provides drinking water for 1.6 million low-income people, aligns with the government's Medium-Term Philippine Development Plan and the Millennium Development Goals. For an overview of Manila Water's sustainablity framework, accessed September 4, 2014, see http://www.manilawater.com/Pages/SustainabilityFramework.aspx.

15. See Manila Water, 2011 Sustainability Report, 67–69, for data on the cooperatives and a more general discussion of Manila Water's engagement with small businesses.

16. From Manila Water's 2012 Annual Report, 15-16: "The construction of three major sewage treatment plants is seen to increase our waste water capacity significantly. These are Taguig North, North Marikina and Ilugin STPS, which have capacities of 75, 100 and 165 mld (million liters per day), respectively. The completion of these facilities will increase our wastewater treatment capacity to more than 450 mld, bringing us closer to our goal of increasing our current 23% wastewater treatment coverage to 63% by 2022."

17. Manila Water, 2009 Annual Report, http://www.manilawater.com/investor/Investor%20Resources/Pages/DownloadableMaterials.aspx (retrieved January 26, 2013).

18. *Good Practices in Urban Water Management*, 104.

19. Manila Water, 2009 Annual Report, 4 (Chairman's Letter).

20. Towers Watson. "Manila Water Company: 10 Questions for Ruel Maranan, Group Director, Corporate Human Resources," January 2014, http://www.towerswatson.com/en/Insights/Newsletters/Global/strategy-at-work/2014/10-questions-for-ruel-maranan. Accessed January 29, 2014. "Manila Water Deploys Mobile Treatment Plant to Quake-Hit Bohol," Interaksyon, October 22, 2013, accessed January 26, 2014, http://www.interaksyon.com/business/73225/manila-water-deploys-mobile-treatment-plant-to-quake-hit-boho.

21. "Manila Water Deploys Mobile Treatment Plant to Quake-Hit Bohol," Interaksyon, October 22, 2013, accessed January 26, 2014, http://www.interaksyon.com/business/73225/manila-water-deploys-mobile-treatment-plant-to-quake-hit-boho.

22. Manila Water, 2009 Annual Report.

23. Ben Arnold O. De Vera, "Business Groups Rally Behind Manila Water, Maynilad in Tiff with MWSS," Interaksyon, July 19, 2013, accessed November 2, 2013, http://www.interaksyon.com/business/66748/business-groups-rally-behind-manila-water-maynilad-in-tiff-with-mwss.

24. Rainfall is 2.34 meters. Cecilia Tortajada, Yugal Joshi, and Asit K. Biswas. *The Singapore Water Story: Sustainable Water Development in an Urban City-State* (Milton Park: Routledge, 2013), 1.

25. The most detailed account of this is to be found in ibid. This book notes that at time of independence consumption was seventy million gallons a day, there were three reservoirs, and the catchment area covered only 11 percent of the country. By 2011, there were seventeen reservoirs and the catchment area comprised two thirds of the country. See ibid., 9 (table).

The "Cut! Cut! Cut!" chant is cited in Beth Duff-Brown, "Malaysia-Singapore Relations Worsen," Associated Press, August 13, 1998, accessed January 29, 2014, http://www.apnewsarchive.com/1998/Malaysia-Singapore-Relations-Worsen/id-f7e82ebefb77d9ec53db1a7390f543e2. The Malaysian-Indonesian alliance is quoted in

Lee Poh Onn, *The Water Issue Between and Singapore and Malaysia: No Solution in Sight?* (Singapore: Institute of Southeast Asian Studies, 2003), 2.

26. Interview with the author, April 16, 2013, Singapore.

27. Lee Kuan Yew built up the Singaporean armed forces in part to guarantee continued access to water. This passage from his memoirs describes a meeting, apparently in 1978 or shortly thereafter, with Dr. Mahathir Mohamad, future Malaysian prime minister:

> He [Mahathir] was direct and asked what we were building the SAF (Singapore Armed Forces) for. I [Lee Kuan Yew] replied equally directly that we feared that at some time or other there could be a random act of madness like cutting off our water supplies, which they [the Malaysians] had publicly threatened whenever there were differences between us. We had not wanted separation. It had been thrust upon us. The Separation Agreement with Malaysia was a part of the terms on which we had left and had been deposited with the United Nations. In this agreement, the Malaysian government had guaranteed our water supply. If this was breached, we would go to the UN Security Council. If water shortage became urgent, in an emergency, we would have to go in, forcibly if need be, to repair damaged pipes and machinery and restore the water flow. I was putting my cards on the table. He denied that any such precipitate action would happen. I said I believe that he would not do this, but we had to be prepared for all contingencies."

Lee Kuan Yew, *Third World to First: The Singapore Story: 1965–2000* (Singapore: Singapore Press Holdings, Marshall Cavendish Editions, 2006), 276.

28. *Innovation in Water* (Singapore PUB) 3 (July 2012): 3. http://www.edb.gov.sg/content/edb/en/industries/industries/environment-and-water.html (retrieved August 24, 2014).

29. Ibid., 9, 14, 16.

30. "Tapping into Singapore's Pool of Resources," Environment & Water Industry Program Office, p. 8, retrieved October 5, 2014, http://edb.gov.sg/content/dam/edb/en/industries/Environment%20and%20Water/downloads/Global%20Hydrohub%20Brochure.pdf.

31. "2007: PUB Singapore," http://www.siwi.org/prizes/stockholmindustrywateraward/winners/pub-singapore/ (retrieved January 26, 2013).

32. *Innovation in Water* (Singapore PUB) 3 (July 2012): 9, http://www.edb.gov.sg/content/edb/en/industries/industries/environment-and-water.html (retrieved February 10, 2013).

33. Chew notes that Australia initially failed to introduce treated sewage water because "they couldn't swing the public perception." Opposition remains, but public attitudes have begun to change as communities struggle with the threat of severe water shortages. In San Diego, more than a decade of opposition to drinking wastewater was overturned in 2012 when a wastewater reclamation plant began operation. Felicity Barringer, "As 'Yuck Factor' Subsides, Treated Wastewater Flows from Taps," *New York Times*, February 9, 2012, accessed January 29, 2014, http://www.nytimes

.com/2012/02/10/science/earth/despite-yuck-factor-treated-wastewater-used-for-drinking.html?pagewanted=all&_r=0; Michael Hopkin, "Recycled Waste Water Gets Go-Ahead for Perth," WAToday, August 1, 2013, accessed May 28, 2014, http://www.watoday.com.au/wa-news/recycled-waste-water-gets-goahead-for-perth-20130801-2r10k.html.

34. *Innovation in Water* (Singapore PUB) 3 (July 2012): 26–27, http://www.edb.gov.sg/content/edb/en/industries/industries/environment-and-water.html (retrieved August 24, 2014).

35. Sembcorp runs the Sembcorp Changi NEWater Plant. Keppel operates the Keppel Seghers Ulu Pandan NEWater Plant. Hyflux operates the Bedok and Seletar NEWater plants.

36. The Hyflux section is based on an interview with Olivia Lum on August 21, 2013, at her office in Singapore and publicly available company material. Company documents include the 2012 Annual Report, http://hyflux.listedcompany.com/misc/ar2012.pdf (accessed November 3, 2013).

37. Hyflux, 2011 Annual Report, http://hyflux.listedcompany.com/misc/ar2011.pdf (accessed March 6, 2014).

38. In 2011, the Pearl River Delta accounted for 26.7 percent of China's total exports. "PRD Economic Profile." Hong Kong Trade Development Council. Accessed October 24, 2014. http://china-trade-research.hktdc.com/business-news/article/Fast-Facts/PRD-Economic-Profile/ff/en/1/1X000000/1X06BW84.htm.

39. Esquel formal interviews included Marjorie Yang, August 2, 2013; Agnes Cheng, December 6, 2012, in Hong Kong; and Agnes Cheng and others during a Gaoming factory visit on January 28, 2013, as well as subsequent e-mail communications, August 21–28, 2014.

40. The last available complete world data are from 2006, when countries averaged about 18.7 percent of water pollution coming from the textile industry. "Water Pollution, Textile Industry (% of total BOD emissions)," World Bank, World Development Indicators, http://data.worldbank.org/indicator/EE.BOD.TXTL.ZS?page=1 (accessed May 28, 2014). *China Daily* cites Ma Jun and IPE, http://www.chinadaily.com.cn/china/2012-10/09/content_15802134.htm.

41. Esquel was lauded, but several others were criticized. http://www.ipe.org.cn/en/alliance/t_detail.aspx?name=%E6%BA%A2%E8%BE%BE.

42. China cannot be accused of having a lax standard. Taiwan has a standard of 100 for the textile industry and up to 160 for some kinds of textile dyeing. http://law.epa.gov.tw/en/laws/480770486.html (accessed February 18, 2013).

43. Barbara Finamore, personal comunication with the author, November 13, 2013 (by e-mail). NRDC is closely involved with the textile industry through its 'Clean By Design' project. See also Institute of Public and Environmental Affairs (IPE) and Natural Resources Defense Council (NRDC). 2014. Greening of the Global Supply Chain. June. Retrieved August 30, 2014. http://www.ipe.org.cn/Upload/IPE-Reports/Report-CITI-EN.pdf. Esquel was ranked third, trailing only Apple and H&M, among the 147 cmpanies on the list, ahead of General Electric, Coca-Cola, Nike and other multinationals (p. 26).

44. Barchi Peleg, "Case GS-48: Esquel Group: A Vertically Integrated Apparel Manufacturer," Stanford Graduate School of Business, 2007 (revised 2013), 16 (exhibit c4).

7. The Tropical Challenge: Saving Asia's Lungs

1. "Green Carbon, Black Trade: Illegal Logging, Tax Fraud and Laundering in the World's Tropical Forests," UNEP and Interpol, p. 6, 2012, accessed May 28, 2014, http://www.unep.org/publications/contents/pub_details_search.asp?ID=6276. Estimates of illegal logging are, not surprisingly, varied; a study commissioned by Chatham House based on 2006 data estimated that the percentage of Indonesia's timber trade that was illegal had fallen from nearly 80 percent in 1998 to 40 percent, citing countrywide efforts implemented in 2005 by the Indonesian government as the reason for the decline. Sam Lawson and Larry MacFaul, *Illegal Logging and Related Trade: Indicators of the Global Response* (London: Chatham House, 2010), 94.

2. United States Environmental Protection Agency. [No Date.] "Global Greenhouse Gas Emissions Data." Accessed August 30, 2014. http://www.epa.gov/climatechange/ghgemissions/global.html. This is based on 2004 data, cited by the Intergovernmental Panel on Climate Change (IPCC) in 2007; as a percentage, the number has almost certainly fallen, given the increase in fossil fuel use, but the fact remains that tropical agriculture is a significant contributor to greenhouse gas emissions.

3. Belinda Arunarwati Margono, Peter V. Potapov, Svetlana Turubanova, Fred Stolle, and Matthew C. Hansen. 2014. *Nature Climate Change*. Volume 4. August. Published Online June 29. "Primary Forest Cover Loss in Indonesia over 2000–2012." Accessed August 30, 2014. http://www.nature.com/nclimate/journal/vaop/ncurrent/full/nclimate2277.html.

4. Siobhan Peters et al., *The Stern Review: Economics of Climate Change* (London: Government of Britain, 2006), 378, http://mudancasclimaticas.cptec.inpe.br/~rmclima/pdfs/destaques/sternreview_report_complete.pdf (accessed May 28, 2014).

5. J. O. Rieley and S. E. Page, "Wise Use of Tropical Peatlands: Focus on Southeast Asia," ALTERRA—Wageningen University and Research Center and the EU INCO—STRAPEAT and RESTORPEAT Partnerships, 2005; Elizabeth Rosenthal, "British Soil Is Battlefield over Peat, for Bogs' Sake," *New York Times*, October 6, 2012, http://www.nytimes.com/2012/10/07/science/earth/british-gardeners-battle-over-peat-for-bogs-sake.html.

6. See, for instance, Ivetta Gerasimchuk and Peng Yam Koh, "The EU Biofuel Policy and Palm Oil: Cutting Subsidies or Cutting Rainforest?" International Institute for Sustainable Development, September 2013, accessed November 30, 2013, http://www.iisd.org/gsi/sites/default/files/bf_eupalmoil.pdf; "Growth in Vegetable Oils Usage for Bio-Fuels Production," Nestlé, April 2010, accessed December 6, 2012, http://www.Nestlé.com/asset-library/Documents/Media/Statements/oil-usage-for-biofuel-production.pdf.

7. "Have a Break?" Greenpeace, YouTube, posted March 10, 2010, accessed March 21, 2014, https://www.youtube.com/watch?v=VaJjPRwExO8.

8. "Progress Report Palm Oil," Nestlé Autumn 2013, accessed August 31, 2014. http://www.nestle.com/asset-library/documents/creating-shared-value/responsible-sourcing/progress-report-palm-oil-autumn-2013.pdf.

9. http://www.Nestlé.com/Media/Statements/Pages/Update-on-deforestation-and-palm-oil.aspx#.UMANd4OTyYg (accessed December 6, 2012); http://www.nestle.com

/asset-library/documents/creating-shared-value/responsible-sourcing/progress-report -palm-oil-autumn-2013.pdf (accessed May 23, 2014). See also "Nestlé Sets Social, Environmental Targets for 2020," Reuters, http://www.reuters.com/article/2013/03/13 /Nestlé-idUSL6N0C563120130313 (accessed April 4, 2013).

10. Unilever met its goal ahead of schedule. "In November 2013, we announced that we would accelerate market transformation towards sustainable palm oil. By the end of 2014, all the palm oil we buy globally will be traceable to known sources i.e. from the originating crude palm oil mills." Unilever. 2014. "Our Sustainable Sourcing Journey," Accessed July 9, 2014. http://www.unilever.com/sustainable-living-2014 /reducing-environmental-impact/sustainable-sourcing/sustainable-palm-oil/.

"Towards Better Practice in Smallholder Palm Oil Production," International Institute for Environment and Development, November 2006, accessed March 20, 2014, http://pubs.iied.org/13533IIED.html.

11. European companies have a long way to go. The WWF publishes an annual scorecard of the palm oil policies of fifty-nine European companies. As of 2013, five companies, including giant retailer Metro, tied for worst, scoring 0. See "2013 Palm Oil Buyers Scorecard," World Wildlife Fund, http://wwf.panda.org/what_we_do /footprint/agriculture/palm_oil/solutions/responsible_purchasing/palm_oil_buyers _scorecard_2013/ (accessed April 1, 2014).

12. Environmentalists don't appear to have easy answers. The WWF says that, in effect, people should stop consuming: "Certification as a means to make the palm oil industry sustainable fails to deal with the root causes of the problem. The destruction caused by the expansion of palm oil is caused by the excessive and irrational use of vegetable oil, either as a foodstuff, industrial oil or agrofuel. Sustainable production can only be achieved by halting the increased demand and over-consumption in order to create sustainable levels of demand. . . ." Oliver Bach, "Sustainable Palm Oil: How Successful Is RSPO Certification?" *The Guardian*, July 4, 2013, accessed April 1, 2014, http://www.theguardian.com/sustainable-business /sustainable-palm-oil-successful-rspo-certification.

13. For the Tahija family, see the autobiographies of Jean and Julius Tahija: Jean Tahija, *An Unconventional Woman* (Ringwood, Victoria: Penguin Books Australia, 1998); Julius Tahija, *Horizon Beyond* (Singapore: Times Editions, 1995).

14. Company and industry data are based on an author interview with Ruslan Krisno, Great Giant Pineapple's Agrigroup sustainability director, February 28, 2013, in Jakarta; a visit to the Lampung, Sumatra, plantation, canning facility, cattle lot, biogas digester, and fertilizer plant, April 2, 2014 featuring interviews with Ruslan Krisno and others; and discussion with Husodo Angkosubroto, February 27, 2013. I also used the company booklet: Great Giant Pineapple. *Earth Water Wind & Fire* (Jakarta: Author, [2010]).

15. Although not directly related to the sustainability efforts, as evidence of the company's long-term approach it is worth noting that Great Giant has an extensive research program that involves aspects such as improving soil health by standardizing fertilizer (including compost) usage, clone development, and best practices, especially in combating disease. About fifty people work in the thirty-hectare research area; the

company cooperates with researchers in Indonesia as well as Taiwan, Germany, and Colombia.

16. "It's too early to give any exact numbers, but we are confident we can make an important contribution to global CO_2 abatement through this program. For example, Indonesian peatland stores 132 gigatons of CO_2. In comparison, the world's largest rainforest, the Amazon, stores 168 gigatons of CO_2. An Indonesian cut of 1.20 gigatons (41 percent reduction) of greenhouse gas emissions by 2020 would equal around 8 percent of the total global reduction required to reach the emissions levels recommended by the United Nations' Intergovernmental Panel on Climate Change and which scientists believe are necessary to prevent the average global temperature from increasing by more than 2 degrees Centigrade."

"Norway-Indonesia REDD+ Partnership-Frequently Asked Questions," Norway, the Official Site in Indonesia, May 31, 2010, http://www.norway.or.id/Norway_in_Indonesia/Environment/-FAQ-Norway-Indonesia-REDD-Partnership-/#.UoNZqaiSzPw. Accessed October. 5, 2014.

17. Ibid.

18. "Is Norway worried that funds will be lost to corruption? Indonesia has a very good track-record of managing foreign donor funds under President Yudhoyono. The Aceh and Nias Rehabilitation and Reconstruction Agency (BRR) established after the 2004 tsunami managed around US$7 billion of foreign donor funds in line with the best international standards. Many of the same governance principles will be used for the special agency that will be established to coordinate the development and implementation of REDD+ in Indonesia. Indonesia and Norway have agreed that the funds will be managed by an internationally reputable financial institution according to international fiduciary, governance, environmental and social standards." (Ibid.)

8. "Adhere and Prosper": One Company's Quest for Green Power

1. https://www.clpgroup.com/ourcompany/news/Pages/18112009.aspx (accessed August 6, 2010).

2. CLP, 2012 Sustainability Report, 36. See also CLP, 2010 Environment Report, 27. However, CLP's carbon-intensity worsened in 2013, as a result of increased coal. CLP. 2014. Sustainability Report. 2013, p. 100. Accessed August 31, 2014. https://www.clpgroup.com/sr2013/en/index.html#100.

3. CLP, 2012 Annual Report, 88:

> Since 2007, we have invested close to US$3 billion in renewable energy assets, aligning our investments with the emergence of policies and incentives across our regional portfolio which have supported zero and low-emissions generation. Our target to have 20 percent of our generating capacity in non-carbon emitting sources by 2020 was met on 31 December 2010, 10 years ahead

of schedule. We have, therefore, raised our non-carbon emitting target to 30 percent by 2020. At the start of 2012, renewable energy accounted for 18.3 percent (equity based) of CLP Group's generation portfolio. To give an idea of the pace of our efforts, in 2004 renewable energy accounted for less than 1 percent of CLP's generation portfolio.

4. CLP, "Our Journey to a Low Carbon Energy Vision," 6.

5. CLP, 2012 Annual Report, 88.

6. Nigel Cameron. 1982. *Power: The Story of China Light.* Hong Kong: Oxford University Press, 37; China Light and Power (CLP). 2001. *A Century of Light.* Hong Kong: China Light and Power, 16.

7. *A Century of Light*, vii (S. Y. Chung letter).

8. Ibid., p. viii.

9. "Castle Peak Power Station," CLP website. https://www.clpgroup.com/ouroperations/assetsandservices/powergeneraton/coalfiredpowerplants/Pages/castlepeakpowerstation.aspx Accessed July 9, 2014. https://www.clpgroup.com/ouroperations/assetsandservices/powergeneraton/coalfiredpowerplants/Pages/castlepeakpowerstation.aspx. Castle Peak was built in the 1980s to end dependence on oil and meet high electricity demand in Hong Kong. CLP, 2012 Annual Report, 42: "In addition, our emissions control project at Castle Peak, that was completed in 2010 and required an investment of around HK$9 billion, has made our plant at Castle Peak one of the cleanest coal-fired power stations in the world in terms of emissions. Combined with many other initiatives, such as the use of low sulphur coal from Indonesia, we have made enormous efforts to reduce emissions."

10. *A Century of Light*, 208. This book also has information on the 1991 Hok Un demolition; Tsing Yi stopped operating in 1996 and was demolished in 1998.

11. Household surveys found that in 1985 average annual net per capita income for rural residents was ¥398, about half the average per capita urban income, which was ¥739. China Statistical Yearbook, 1999. Beijing: National Bureau of Statistics of the People's Republic of China, 1999. Accessed October 23, 2014. http://www.stats.gov.cn/yearbook/indexC.htm

12. According to Lawrence Kadoorie, "This projected station would be the *first* nuclear power station in China." Cameron, op. cit., p., 261.

13. For the Yangjiang announcement, see CLP. 2013. "Announcement Concerning Yangjiang Nuclear Investment." September 3. Retrieved September 1, 2014, https://www.clpgroup.com/ourcompany/aboutus/resourcecorner/investmentresources/Announcement%20%20Circulars/2013/e_Announcement%20concerning%20Yangjiang%20nuclear%20investment%20(20130903).pdf.

14. CLP, "Climate Vision 2050 Announcement," https://www.clpgroup.com/ourcompany/news/Pages/07122007.aspx.

15. CLP, 2012 Annual Report, 7 (Chairman's Statement), https://www.clpgroup.com/ourcompany/news/Pages/15112007.aspx.

Bibliography

Ali, Saleem. 2012. "The Promise of Chinese Eco-cities." National Geographic News-Watch. November 23. Accessed March 9, 2014. http://newswatch.nationalgeographic.com/2012/11/23/eco-cities/.

American Institute of Architects. Toolkit 2030. Accessed August 29, 2014. https://info.aia.org/toolkit2030/advocacy/architects-green-building.html.

Alliance to Save Energy. "Our History." Accessed August 15, 2013. http://www.ase.org/about-us/our-history.

——. 2014. "Post-Event." October 3. Accessed May 29, 2014. http://www.ase.org/events/2013-evening-stars-energy-efficiency-awards-dinner.

Araki, Laura. 2012. "Fukushima One Year Later: An Interview with Daniel P. Aldrich." National Bureau of Asian Research. March 6. Accessed February 4, 2014. http://www.nbr.org/research/activity.aspx?id=219#.UxsBe9zzZyM.

Asia Business Council. "CII Sohrabji Godrej Green Business Center." Accessed August 16, 2013. http://www.asiabusinesscouncil.org/docs/BEE/GBCS/GBCS_CII.pdf.

Asian Development Bank. 2012. "Asia's Shift to Greener Transport Key to Global Sustainability—ADB Forum." Asian Development Bank News and Events. November 6. Accessed May 28, 2014. http://www.adb.org/news/asias-shift-greener-transport-key-global-sustainability-adb-forum.

——. 2013. *Asian Development Outlook 2013: Asia's Energy Challenge.* Manila: Asian Development Bank.

ASVT Coalition. 2009. "Toward a Just and Sustainable Solar Industry: A Silicon Valley Toxics Coalition Whitepaper." January 14. Accessed May 26, 2014. http://svtc.org/wp-content/uploads/Silicon_Valley_Toxics_Coalition_-_Toward_a_Just_and_Sust.pdf.

Austindo Nusantara Jaya Tbk. 2013. Preliminary Share Offering Memorandum. April 15. Sole Global Coordinator, International Selling Agent, and Bookrunner: Morgan Stanley.

Australia, Government of, Treasury. 2011. "Chinese Macroeconomic Management Through the Crisis and Beyond." Accessed July 1, 2013. http://www.treasury.gov.au/PublicationsAndMedia/Publications/2011/Chinese-Macroeconomic-Management-Through-the-Crisis-and-Beyond/working-paper-2011–01/Chinas-stimulus-package.

Bach, Oliver. "Sustainable Palm Oil: How Successful Is RSPO Certification?" *The Guardian*. July 4. Accessed April 1, 2014. http://www.theguardian.com/sustainable-business/sustainable-palm-oil-successful-rspo-certification.

Ball, Jeffrey. 2013. "China's Solar-Panel Boom and Bust." Stanford Graduate School of Business. June 7. Accessed May 16, 2014. https://www.gsb.stanford.edu/news/headlines/chinas-solar-panel-boom-bust.

Bansal, Sarika, and Skylar Bergl. 2013. "Most Innovative Companies: Mahindra Reva, Tesla." *Fast Company*. Accessed June 22, 2103. http://www.fastcompany.com/most-innovative-companies/2013/mahindra-reva-tesla.

Barringer, Felicity. 2012. "As 'Yuck Factor' Subsides, Treated Wastewater Flows from Taps." *New York Times*. February 9. Accessed January 29, 2014. http://www.nytimes.com/2012/02/10/science/earth/despite-yuck-factor-treated-wastewater-used-for-drinking.html?pagewanted=all&_r=0.

Bartlett, John, ed. 2005 [e-book]. *Familiar Quotations*. Accessed June 6, 2014. http://www.gutenberg.org/files/16732/16732-h/16732-h.htm.

BBC. 2013. "Indonesia Fuel Prices Rocket by 44% Sparking Protests." June 22. Retrieved July 1, 2013. http://www.bbc.co.uk/news/world-asia-23015511.

BBC News. "Horizons: Smart Cities." Accessed March 9, 2014. http://www.bbc.com/specialfeatures/horizonsbusiness/.

Bell, Michelle L., Devra L. Davis, and Tony Fletcher. 2004. "A Retrospective Assessment of Mortality from the London Smog Episode of 1952: The Role of Influenza and Pollution." *Environmental Health Perspectives* 112, no. 1: x.

Belson, Ken. 2008. "Importing a Decongestant for Midtown Streets." *New York Times*. March 16. Accessed August 4, 2013. http://www.nytimes.com/2008/03/16/automobiles/16CONGEST.html?ex=1363320000&en=66db1c235736c7ea&ei=5088&partner=rssnyt&emc=rss.

Benfield, Kaid. 2013. "As Good and Important as It Is, LEED Can Be So Embarrassing." Switchboard: Natural Resources Defense Council. January 18. Accessed June 7, 2014. http://switchboard.nrdc.org/blogs/kbenfield/as_good_and_important_as_it_is.html.

Blinch, Jenny, Benjamin McCarron, Katie Yewdall, and Lucy Carmody. 2011. *The Future of Fish in Asia*. Singapore: Responsible Research.

Bloomberg Businessweek. Xinjiang Goldwind Sci&Tech-A (002202: Shenzhen). Accessed August 23, 2014. http://investing.businessweek.com/research/stocks/people/person.asp?personId=37616794&ticker=002202:CH.

Bloomberg New Energy Finance. 2012. "Overcapacity and New Players Keep Wind Turbine Prices in the Doldrums." Press release. March 6. Accessed June 7, 2014. http://about.bnef.com/press-releases/overcapacity-and-new-players-keep-wind-turbine-prices-in-the-doldrums/.

——. 2014. "China's 12GW Solar Market Outstripped All Expectations in 2013." January 23. Accessed March 6, 2014. http://about.bnef.com/press-releases/chinas-12gw-solar-market-outstripped-all-expectations-in-2013/.

——. 2014. "China Out-spends the US for First Time in $15bn Smart Grid Market." February 18. Accessed March 6, 2014. http://about.bnef.com/press-releases/china-out-spends-the-us-for-first-time-in-15bn-smart-grid-market/.

Bloomberg News. 2013. "Solar Defaults Shock Holders as $8.4 Billion Due: China Credit." September 3. Accessed May 26, 2013. http://www.bloomberg.com/news/2013–09–03/solar-defaults-shock-holders-as-8–4-billion-due-china-credit.html.

——. 2013. "Chinese Zombies Emerging After Years of Solar Subsidies." September 9. Accessed March 6. 2014. http://www.bloomberg.com/news/2013–09–08/chinese-zombies-emerging-after-years-of-solar-subsidies.html.

——. 2013. "China's Limit on New Solar Factories Seen Driving M&A." September 18. Accessed March 8, 2014. http://www.bloomberg.com/news/2013–09–17/china-to-strictly-limit-building-of-more-photovoltaic-capacity.html.

——. 2014. "Chinese Wind Turbine Makers Face Market Consolidation." April 18. Accessed August 28, 2014. http://www.bloomberg.com/news/2014-04-17/china-s-wind-turbine-makers-face-consolidation-as-glut-lingers.html.

——. 2014. "China Adds Australia-Sized Capacity in Energy Push." August 7. Accessed August 29, 2014. http://www.bloomberg.com/news/2014-08-07/china-add-australia-sized-solar-capacity-in-energy-push.html.

Boston-Power. 2011. "Boston-Power Announces $125 Million in New Financing." September 20. Accessed August 30, 2014. http://www.boston-power.com/news/press-releases/boston-power-announces-125-million-new-financing.

BP. 2014. BP Energy Outlook 2035. January. Accessed August 31, 2014. http://www.bp.com/content/dam/bp/pdf/Energy-economics/Energy-Outlook/Energy_Outlook_2035_booklet.pdf.

Bradsher, Keith. 2012. "China Opens Longest High-Speed Rail Line." *New York Times*. December 26. Retrieved April 11, 2014. http://www.nytimes.com/2012/12/27/business/global/worlds-longest-high-speed-rail-line-opens-in-china.html.

Buczynski, Beth. 2013. "7 Most Impressive Wind Farms (and Turbines) in the World." Global Wind Energy Council, November 6. Accessed November 10, 2013. http://www.care2.com/causes/7-most-impressive-wind-farms-and-turbines-in-the-world.html.

BusinessWeek. 1999. "Singapore's Lee: 'You Shape Up or Perish.'" June 20. Accessed August 4, 2013. http://www.businessweek.com/stories/1999–06–20/singapores-lee-you-shape-up-or-perish-extended.

——. 2003. "Wang Chuanfu." June 8. Accessed April 7, 2014. http://www.businessweek.com/stories/2003-06-08/wang-chuanfu.

——. "China's Most Powerful People 2009: Wu Gang," accessed November 15, 2013, http://images.businessweek.com/ss/09/11/1113_business_stars_of_china/28.htm.

BusinessWire. 2010. "BYD Tops Bloomberg BusinessWeek's 12th Annual Tech 100 List." May 20. Retrieved July 20, 2013. http://www.businesswire.com/news/home/20100520006751/en/BYD-Tops-Bloomberg-Businessweek%E2%80%99s-12th-Annual-Tech.

——. 2012. "Polysilicon Prices Hit Record Low in 2011; Will Head Even Lower Enabling $0.70 per Watt Solar Panels in 2012." Press release. January 19. Accessed March 8, 2014. http://www.businesswire.com/news/home/20120119005221/en/Polysilicon-Prices-Hit-Record-2011-Head-Enabling.

BYD. Annual Report 2010. Accessed April 11, 2014. http://bydit.com/doce/investor/CorporateReport/AnnualReport/.

——. Annual Report 2011. Accessed April 11, 2014. http://bydit.com/doce/investor/CorporateReport/AnnualReport/.

——. Annual Report 2012. Accessed August 31, 2014. http://bydit.com/userfiles/attachment/ANNUAL%20REPORT%202012%EF%BC%88PRINTED%20VERSION%EF%BC%89.pdf.

——. Annual Report 2013. Accessed August 31, 2014. http://bydit.com/userfiles/attachment/20140429-BYD%20COMPANY%20LIMITED%20%20ANNUAL%20REPORT%202013.pdf.

——. "Company Profile." Accessed April 7, 2014. http://www.byd-auto.net/company/profile.php.

California Environmental Protection Agency. 2013. "Zero-Emission Vehicle Legal and Regulatory Activities and Background." June 10. Accessed July 17, 2013. http://www.arb.ca.gov/msprog/zevprog/zevregs/zevregs.htm.

Cameron, Nigel. 1982. *Power: The Story of China Light.* Hong Kong: Oxford University Press.

Capital Market. 2000. "Top Rankers in 2000." December 25. Accessed April 8, 2010. http://www.capitalmarket.com/magazine/cm1521/covsto.htm.

Center for Climate and Energy Solutions. "Coal." Accessed August 28, 2014. http://www.c2es.org/energy/source/coal.

Center for Liveable Cities. 2013. "Report: From Garden City to City in a Garden" (Singapore Lecture Series). July 15. http://www.clc.gov.sg/documents/Lectures/2013/Greening_Singapore_report.pdf.

Chan, Stephen. 2012. "Stop Tall Building Says Party Paper." *South China Morning Post.* August 13, A6.

Chellaney, Brahma. 2011. *Water: Asia's New Battleground.* Washington, DC: Georgetown University Press.

——. 2013. *Water, Peace, and War: Confronting the Global Water Crisis.* Lanham, MD: Rowman & Littlefield.

ChinaAutoWeb. 2014. "Plug-In EV Sales in China Rose 37.9% to 17,600 in 2013." January 10. http://chinaautoweb.com/2014/01/plug-in-ev-sales-in-china-rose-37-9-to-17600-in-2013/.

China Labor Bulletin. 2008. "Bone and Blood: The Price of Coal in China," CLB Research Report No. 6, March. Accessed August 28, 2014. http://www.clb.org.hk/en/files/File/bone_and_blood.pdf.

——. 2014. "Coal Mine Accidents Decrease as Production Stagnates." April 3. Accessed August 28, 2014. http://www.clb.org.hk/en/content/coal-mine-accidents-china-decrease-production-stagnates.

China Light and Power (CLP). 2001. *A Century of Light.* Hong Kong: China Light and Power.

——. 2010. *Our Journey to a Low Carbon Energy Vision.* Hong Kong: China Light and Power.

——. 2013. 2012 Annual Report. Hong Kong: China Light and Power. https://www.clpgroup.com/ourcompany/news/Pages/15112007.aspx.

——. 2013. 2012 Sustainability Report. Hong Kong: China Light and Power.

——. 2007. "Climate Vision 2050 Announcement." https://www.clpgroup.com/ourcompany/news/Pages/07122007.aspx.

——. 2013. "Announcement Concerning Yangjiang Nuclear Investment." September 3. Accessed September 1, 2014. https://www.clpgroup.com/ourcompany/aboutus/resourcecorner/investmentresources/Announcement%20%20Circulars/2013/e_Announcement%20concerning%20Yangjiang%20nuclear%20investment%20(20130903).pdf.

——. 2014. "Castle Peak Power Station," CLP website. https://www.clpgroup.com/ouroperations/assetsandservices/powergeneraton/coalfiredpowerplants/Pages/castlepeakpowerstation.aspx Accessed July 9, 2014.

China Merchants New Energy Group. 2012. "Group Resources." Accessed May 7, 2014. http://www.cmnechina.com/index.php?langid=en.

Chiplunkar, Anand, et al., eds. 2012. *Good Practices in Urban Water Management: Decoding Good Practices for a Successful Future.* Manila: Asian Development Bank. http://www.adb.org/sites/default/files/pub/2012/good-practices-urban-water-management.pdf.

CIA World Factbook. "Country Comparison: Electricity: Installed Generating Capacity." Accessed August 23, 2014. https://www.cia.gov/library/publications/the-world-factbook/rankorder/2236rank.html.

Clifford, Mark L. 2010. "Korea's Green Revolution." In *Korea 2020: Imagining the Next Decade.* Seoul: Random House Korea.

—— and Janet Pau. 2012. *Through the Eyes of Tiger Cubs: Views of Asia's Next Generation.* Singapore: John Wiley & Sons (Asia).

Cobb, Jeff. 2011. "LG Chem Opens 'Largest' EV Battery Plant." Gm-Volt.com. April 15. Accessed April 4, 2014. http://gm-volt.com/2011/04/15/lg-chem-opens-ochong-battery-plant-expects-major-market-share/.

Cox, Wendell. 2012. "World Urban Areas Population and Density: A 2012 Update." NewGeography. May 3. Accessed July 22, 2013. http://www.newgeography.com/content/002808-world-urban-areas-population-and-density-a-2012-update.

Czuczka, Tony. 2012. "Germany Cuts Solar Aid to Curb Prices, Panel Installations." Bloomberg News. March 29. Accessed March 6, 2014. http://www.businessweek.com/news/2012–03–29/germany-cuts-solar-aid-to-curb-prices-panel-installations.

Dai Qing. 1998. *The River Dragon Has Come!* Armonk, NY: M. E. Sharpe.

Davison, Nicola. 2013. "Is the Shanghai Tower the World's First Eco-friendly Skyscraper?" China Dialogue. October 14. Accessed June 2, 2014. https://www.chinadialogue.net/article/show/single/en/6413-Is-the-Shanghai-Tower-the-world-s-first-eco-friendly-skyscraper-.

De Vera, Ben Arnold O. 2013. "Business Groups Rally Behind Manila Water, Maynilad in Tiff with MWSS." InterAksyon.com. July 19. Accessed November 2, 2013. http://www.interaksyon.com/business/66748/business-groups-rally-behind-manila-water-maynilad-in-tiff-with-mwss.

Dobbs, Richard, Sven Smit, Jaana Remes, James Manyika, Charles Roxburgh, and Alejandra Restrepo. 2011. *Urban World: Mapping the Economic Power of Cities.* McKinsey Global Institute.

Dodson, Bill. 2012. *China Fast Forward: The Technologies, Green Industries and Innovations Driving the Mainland's Future.* Singapore: John Wiley & Sons Singapore.

Duff-Brown, Beth. 1998. "Malaysia-Singapore Relations Worsen." Associated Press. August 13. Accessed January 29, 2014. http://www.apnewsarchive.com/1998/Malaysia-Singapore-Relations-Worsen/id-f7e82ebefb77d9ec53db1a7390f543e2.

Economy, Elizabeth. 2010. *The River Runs Black: The Environmental Challenges to China's Future.* 2nd ed. Ithaca, NY: Cornell University Press (A Council on Foreign Relations Book).

——. 2014. "China File: How Responsible Are Americans for China's Pollution Problem?" Council on Foreign Relations. February 28. Accessed March 1, 2014. http://www.cfr.org/china/china-file-responsible-americans-chinas-pollution-problem/p32495.

——. 2014. "China Wakes Up to Its Environmental Catastrophe." Bloomberg Businessweek. March 13. Accessed May 24, 2014. http://www.businessweek.com/articles/2014–03–13/china-wakes-up-to-its-environmental-catastrophe#p1.

Elledge, Jonn. 2014. "The Most Polluted Cities on Earth Are Not Where You Think." New Statesman. May 16. Accessed August 31, 2014. http://www.newstatesman.com/jonn-elledge/2014/05/most-polluted-cities-earth-are-not-where-you-think.

Emanuel, Kerry A. "Downscaling CMIP5 Climate Models Shows Increased Tropical Cyclone Activity over the 21st Century." *Proceedings of the National Academy of Sciences of the United States of America.* Approved [for publication] June 10, 2013. Accessed May 1, 2014. http://www.pnas.org/content/early/2013/07/05/1301293110.abstract?sid=9fb226fc-6f82–4b7a-8c91–6ce889da1b1a.

Esquel Group. 2008. *30th Anniversary History of Esquel.* 3 vols. Hong Kong: Esquel Group.

European Wind Energy Association. 2014. "Statistics." February. Accessed May 23, 2014. http://www.ewea.org/statistics/.

Fabi, Randy and Wilda Asmarini. 2014. "Economists Back Jokowi's Fuel Subsidy Plan Cut for Indonesia."Reuters. May 12. Accessed August 30, 2014. http://www.reuters.com/article/2014/05/12/us-indonesia-election-budget-idUSBREA4B03V20140512.

Fallows, James. 2008. *Postcards from Tomorrow Square: Reports from China.* New York: Vintage Books.

Fast Company. 2013. "Most Innovative Companies 2010: BYD." http://www.fastcompany.com/mic/2010/profile/byd.

Finamore, Barbara. 2013. "A Tale of Two Cities: Energy Efficient Buildings in New York and Hong Kong." Switchboard: Natural Resources Defense Council. May 14. Accessed June 6, 2014. http://switchboard.nrdc.org/blogs/bfinamore/a_tale_of_two_cities_energy_ef.html.

Financial Times. "Siam Cement PCL." Accessed April 4, 2014. http://markets.ft.com/research/Markets/Tearsheets/Financials?s=SCC:SET&subview=IncomeStatement.

Flinn, Ryan. 2013. "California Researchers Detect Radioactive Release from Fukushima Plant." Bloomberg News. August 16. Accessed March 5, 2014. http://www.bloomberg.com/news/2011-08-15/california-researchers-detect-radioactive-release-from-fukushima-plant.html.

Flint, Sunshine. 2013. "Living in: Cities of the Future." BBC Travel. June 19. Accessed August 10, 2013. http://www.songdo.com/Uploads/FileManager/BBC%20Travel%206.19.2013.pdf.

Forbes. 2013. "GuruFocus: Buffett's BYD Profit Up to 200% as First Quarter 2013 Profit Triples." April 26. Accessed April 7, 2014. http://www.forbes.com/sites/gurufocus/2013/04/26/buffetts-byd-profit-up-to-200-as-first-quarter-2013-profit-triples/.

Fox, Robin Lane. 2013. "Worth the Gamble." *Financial Times*. March 29. Accessed August 4, 2013. http://www.ft.com/intl/cms/s/2/98778d74-914b-11e2-b4c9-00144feabdc0.html#axzz2azQPpsfp.

Frankel, David, Kenneth Ostrowski, and Dickon Pinner. 2014. "The Disruptive Potential of Solar Power." *McKinsey Quarterly* no. 2: 50–55.

Fung, Justin, et al. 2009. "From Gray to Green: How Energy Efficient Buildings Can Help Make China's Rapid Urbanization Sustainable." BCG and NRDC. October. Accessed May 30, 2014. http://www.bcg.com.cn/en/files/publications/reports_pdf/BCG_From_Gray_to_Green_Oct_2009_tcm42-32257x1x-website.pdf.

Gensler. 2013. "Tallest Building in China Tops Out: Shanghai Tower Completes Historical Ascension." Press release. August 2. Accessed March 13, 2014. http://m.gensler.com/about-us/press-release/shanghai-tower-tops-out.

——. 2013. "Shanghai Tower: Gensler Tops Out China's Tallest Building in 2013." Accessed June 2, 2014. http://du.gensler.com/vol5/shanghai-tower/#/building-facts.

George, Timothy S. 2001. *Minamata: Pollution and the Struggle for Democracy in Postwar Japan*. Cambridge: Harvard University Asia Center.

Gerasimchuk, Ivetta, and Peng Yam Koh. 2013. "The EU Biofuel Policy and Palm Oil: Cutting Subsidies or Cutting Rainforest?" International Institute for Sustainable Development. http://www.iisd.org/gsi/sites/default/files/bf_eupalmoil.pdf.

Glaeser, Edward. 2011. *Triumph of the City: How Our Greatest Invention Makes Us Richer, Smarter, Greener, Healthier and Happier*. London: Macmillan.

Global Environment Facility. 2012. "Investing in Sustainable Transport and Urban Systems." December. http://www.thegef.org/gef/sites/thegef.org/files/publication/gef_transportBrch_nov2012_r3.pdf.

Global Wind Energy Council. 2011. "Global Wind Report: Annual Market Update 2010." Accessed August 28, 2014. http://gwec.net/wp-content/uploads/2012/06/GWEC_annual_market_update_2010_-_2nd_edition_April_2011.pdf.

——. 2012. "Global Wind Report: Annual Market Update 2011." Accessed August 28, 2014, http://gwec.net/wp-content/uploads/2012/06/Annual_report_2011_lowres.pdf.

——. 2013. "Global Wind Report: Annual Market Update 2012." Accessed March 8, 2014. http://www.gwec.net/wp-content/uploads/2012/06/Annual_report_2012_LowRes.pdf.

——. 2014. "Global Wind Statistics 2013." Accessed May 23, 2014. http://www.gwec.net/wp-content/uploads/2014/02/GWEC-PRstats-2013_EN.pdf.

——. 2014. "Global Wind Report: Annual Market Update 2013." Accessed August 28, 2014. http://www.gwec.net/wp-content/uploads/2014/04/GWEC-Global-Wind-Report_9-April-2014.pdf.

Global Wind Energy Council and Greenpeace. 2012. "Global Wind Energy Outlook 2012." Accessed March 8, 2014. http://www.gwec.net/wp-content/uploads/2012/11/GWEO_2012_lowRes.pdf.

Goldwind (Xinjiang Goldwind Science & Technology). 2011. Annual Report 2010. Accessed March 9, 2014. http://www.goldwindglobal.com/web/investor.do.

——. 2012. Annual Report 2011. Accessed August 23, 2014. http://www.goldwindglobal.com/web/investor.do.

——. 2013. Annual Report 2012. Accessed August 23, 2014. http://www.goldwindglobal.com/web/investor.do.

——. 2014. Annual Report 2013. Accessed August 23, 2014. http://www.goldwindglobal.com/web/investor.do.

Goossens, Ehren, and Benjamin Haas. 2014. "Mystery Property Tycoon Makes $533 Million Bet on Solar." April 2. Bloomberg. Accessed August 31, 2014. http://www.bloomberg.com/news/2014-03-31/hong-kong-property-tycoon-makes-533-million-bet-on-solar.html.

Great Giant Pineapple. [2010]. *Earth Water Wind & Fire.* Jakarta: Great Giant Pineapple.

Greenpeace. 2010. "Have a Break?" YouTube. Posted March 10. Accessed March 21, 2014. https://www.youtube.com/watch?v=VaJjPRwExO8.

Halper, Mark. 2012. "Japan, S. Korea Dominate Electric Car and Hybrid Battery Market." Smartplanet. February 15. Accessed May 29, 2014. http://www.smartplanet.com/blog/intelligent-energy/japan-s-korea-dominate-electric-car-and-hybrid-battery-market/.

Hang Lung Properties. 2013. Sustainability Report 2012. Accessed May 29, 2014. http://www.hanglung.com/Libraries/Document_-_Sustainability/Sustainability_Report_2012.sflb.ashx.

Han Qiao. 2013. "China's High-Speed Programme Back on Track." *International Railway Journal.* January 10. Accessed May 18, 2013. http://www.railjournal.com/index.php/high-speed/chinas-high-speed-programme-back-on-track.html.

Hawkins, T. R., B. Singh, G. Majeau-Bettez, and A. H. Strømman. 2013. "Comparative Environmental Life Cycle Assessment of Conventional and Electric Vehicles." *Journal of Industrial Ecology* 17: 53–64.

Health Effects Institute. 2013. "Ambient Air Pollution Among Top Global Health Risks in 2010." March 31. Accessed May 24, 2014. http://www.healtheffects.org/International/HEI-GBD-MethodsSummary-033113.pdf.

Hilton, Isabel. 2009. "Greening Infosys." China Dialogue. October 14. Accessed July 10, 2013. http://www.chinadialogue.net/article/show/single/en/3285-Greening-Infosys.

Ho, Elaine Lynn-Ee, Chih Yuan Woon, and Kamalini Ramdas. 2013. *Changing Landscapes of Singapore*. Singapore: National University of Singapore.

Hong Kong, Government of. 2010. "Hong Kong's Climate Change Strategy and Action Agenda." Environment Bureau. September. Accessed March 9, 2014. http://www.susdev.gov.hk/html/en/council/Paper07–10Annexe.pdf.

——. 2011. "South Island Line (East) and Kwun Tong Line Extension to Commence Works Soon." May 18. Accessed April 8, 2014. http://www.info.gov.hk/gia/general/201105/18/P201105180326.htm.

Hong Kong Trade Development Council. March 25, 2014. "PRD Economic Profile." Accessed October 24, 2014. http://china-trade-research.hktdc.com/business-news/article/Fast-Facts/PRD-Economic-Profile/ff/en/1/1X000000/1X06BW84.htm.

Hopkin, Michael. 2013. "Recycled Waste Water Gets Go-Ahead for Perth." WAToday. August 1. Accessed February 8, 2014. http://money.cnn.com/magazines/fortune/fortune_archive/2006/03/06/8370702/.

Huang Qili. 2013. "The Development Strategy for Coal-Fired Power Generation in China." *Cornerstone*. June 4. Accessed March 8, 2014. http://cornerstonemag.net/the-development-strategy-for-coal-fired-power-generation-in-china/.

Huet, Natalie, and Caroline Copley. 2014. "Holcim, Lafarge Agree to Merger to Create Cement Giant." Reuters. April 8. Accessed May 28, 2014. http://in.reuters.com/article/2014/04/07/lafarge-holcim-idINDEEA3607P20140407.

Hyflux. 2012. 2011 Annual Report. Accessed March 6, 2014. http://hyflux.listedcompany.com/misc/ar2011.pdf.

——. 2013. 2012 Annual Report. Accessed November 3, 2013. http://hyflux.listedcompany.com/misc/ar2012.pdf.

Hyundai Motor America. 2014. "Hyundai Fuel Cell Press Releases." Accessed August 30, 2014. http://www.hyundainews.com/us/en-us/FuelCell/PressReleases.aspx.

Hyunjoo, Jin, and Christian Hetzner. 2012. "Hyundai Revs Up Fuel-Cell Plan as Battery Technology Disappoints." Reuters. September 25. Accessed April 7, 2014. http://www.reuters.com/article/2012/09/25/us-autoshow-paris-hyundai-fuelcell.

India Green Building Council. "Indian Green Building Council Scorecard." Accessed March 8, 2014. http://igbc.in/site/igbc/index.jsp.

Information Office of the State Council. 2012. "Full Text: China's Energy Policy 2012." Xinhua. October 24. Accessed November 23, 2013. http://news.xinhuanet.com/english/china/2012–10/24/c_131927649.htm.

Infosys. 2012. Sustainability Report, 2011–12. Dowloaded August 29, 2014. http://www.infosys.com/sustainability/Documents/infosys-sustainability-report-2011-12.pdf.

——. 2013. Sustainability Report, 2012–13. Downloaded February 7, 2014. http://www.infosys.com/sustainability/Documents/infosys-sustainability-report-2012–13.pdf.

——. 2014. Infosys Sustainability Report, 2013-14. Downloaded August 29, 2014. http://www.infosys.com/sustainability/documents/infosys-sustainability-report-2013-14.pdf.

Institutional Investors Group on Climate Change. 2013. "Protecting Investment Value in Real Estate: Managing Investment Risks from Climate Change." Accessed August 31, 2014. http://www.iigcc.org/publications/publication/protecting-value-in-real-estate-managing-investment-risks-from-climate-chan.

Interaksyon. 2013. "Manila Water Deploys Mobile Treatment Plant to Quake-Hit Bohol." October 22. Accessed January 26, 2014. http://www.interaksyon.com/business/73225/manila-water-deploys-mobile-treatment-plant-to-quake-hit-boho.

International Energy Agency. 2013. "Trends 2013 in Photovoltaic Applications." November 29. Accessed March 5, 2013. http://www.iea-pvps.org/index.php?id=3&eID=dam_frontend_push&docID=1733.

——. 2013. "Technology Road Map: Wind Energy, 2013 Edition." Accessed March 8, 2014. http://www.iea.org/publications/freepublications/publication/Wind_2013_Roadmap.pdf.

——. "Southeast Asia Energy Outlook." Accessed June 6, 2014. http://www.iea.org/publications/freepublications/publication/SoutheastAsiaEnergyOutlook_WEO2013SpecialReport.pdf.

——. "FAQs: Energy Efficiency." Accessed August 29, 2014. http://www.iea.org/aboutus/faqs/energyefficiency/.

International Institute for Environment and Development. 2006. "Towards Better Practice in Smallholder Palm Oil Production." November. Accessed March 20, 2014. http://pubs.iied.org/13533IIED.html.

International Monetary Fund. 2013. "IMF Calls for Global Reform of Energy Subsidies: Sees Major Gains for Economic Growth and the Environment." Press release. Accessed July 20, 2013. http://www.imf.org/external/np/sec/pr/2013/pr1393.htm.

International Organization of Motor Vehicle Manufacturers. "New PC Registrations or Sales." Accessed May 28, 2014. http://www.oica.net/wp-content/uploads//pc-sales-2013.pdf.

The Institute of Public and Environmental Affairs, Lvse Jiangnan Public Environmental Concerned Center, Green Hunan, and SynTao. 2013. "Responsible Investment in the Cement Industry: Still a Long Way to Go." June 18. Accessed August 31, 2014. http://www.ipe.org.cn/Upload/IPE-Reports/Report-Cement-Phase-I-EN.pdf.

The Institute of Public and Environmental Affairs (IPE) and Natural Resources Defense Council (NRDC). 2014. Greening of the Global Supply Chain. June. Accessed August 30, 2014. http://www.ipe.org.cn/Upload/IPE-Reports/Report-CITI-EN.pdf.

Invest India. "Cement Sector." Accessed August 14, 2013. http://www.investindia.gov.in/?q=cement-sector.

IRIN News. 2010. "ASIA: Top 10 Deadliest Cyclones." September 23. Accessed October 22, 2013. http://www.irinnews.org/report/90556/asia-top-10-deadliest-cyclones.

Jäger-Waldau, Arnulf. 2013. "PV Status Report 2013." European Commission. September. Accessed March 8, 2014. http://iet.jrc.ec.europa.eu/remea/sites/remea/files/pv_status_report_2013.pdf.

Junfeng, Li, et al. 2012. "China Wind Energy Outlook 2012." Accessed March 9, 2014. http://www.gwec.net/wp-content/uploads/2012/11/China-Outlook-2012-EN.pdf.

Kaiman, Jonathan. 2014. "China's Toxic Air Pollution Resembles Nuclear Winter, Say Scientists." *The Guardian*. February 25. Accessed May 1, 2014, http://www.theguardian.com/world/2014/feb/25/china-toxic-air-pollution-nuclear-winterscientists?CMP=fb_gu.

Karmini, Niniek. 2013. "Indonesia Steps Toward Fuel Hike Despite Protests." Associated Press. June 17. Accessed April 4, 2014. http://bigstory.ap.org/article/indonesia-expected-approve-fuel-hike.

Kasarda, John D., and Greg Lindsay. 2011. *Aerotropolis: The Way We'll Live Next*. New York: Farrar, Straus and Giroux.

Keppel Seghers. Accessed February 11, 2014. http://www.keppelseghers.com/en/.

King, Ed, Responding to Climate Change (RTCC). 2014. "Is China Really Committed to Addressing Climate Change?" January 10. Accessed August 28, 2014. http://www.rtcc.org/2014/01/10/is-china-really-committed-to-addressing-climate-change/.

King, Ritchie, and Lily Kuo. 2013. "Here Are the World's Worst Cities for Air Pollution, and They're Not the Ones You'd Expect." Quartz. October 18. Accessed March 5, 2014. http://qz.com/136606/here-are-the-worlds-worst-cities-for-air-pollution-and-theyre-not-the-ones-youd-expect/.

Kingston, Jeff. 2012. "Japan's Nuclear Village." *Asia-Pacific Journal* 10, no. 1 (September 10). Accessed February 4, 2014. http://www.japanfocus.org/-Jeff-Kingston/3822#sthash.h5P4m1Le.dpuf.

Kirkland, Joel. 2011. "China's Ambitious, High-Growth 5-Year Plan Stirs a Climate Debate." *New York Times* (reprint of story from Climate Wire). April 12. Accessed June 7, 2014. http://www.nytimes.com/cwire/2011/04/12/12climatewire-chinas-ambitious-high-growth-5-year-plan-sti-12439.html?pagewanted=all.

Kishnani, Nirmal. 2012. *Greening Asia: Emerging Principles for Sustainable Architecture*. Singapore: BCI Asia.

KPMG International Cooperative. "Global Tax Rates Table." Accessed October 8, 2013. http://www.kpmg.com/global/en/services/tax/tax-tools-and-resources/pages/corporate-tax-rates-table.aspx.

Krieger, Axel, Philipp Radtke, and Larry Wang. 2012. "Recharging China's Electric Vehicle Aspirations." July. Accessed July 20, 2013. http://www.mckinsey.com/insights/energy_resources_materials/recharging_chinas_electric-vehicle_aspirations.

Kwok Siu-tong. 2001. *A Century of Light*. Hong Kong: CLP Power Hong Kong.

Kynge, James. 2007. *China Shakes the World: A Titan's Rise and Troubled Future—and the Challenge for America*. New York: Mariner Books.

Lafarge. 2012. *Sustainability Ambitions 2020: Figures and Objectives*. Paris: Lafarge.

Lafarge India Construction Development Lab. 2013. *Bring Innovation Closer to Markets Through Construction System Developments*. Mumbai: Lafarge India.

Lang, Ng. 2008. "A City in a Garden." World Cities Summit Issue. Singapore: Civil Service College. Accessed August 4. 2013. http://www.cscollege.gov.sg/Knowledge/Ethos/World%20Cities%20Summit/Pages/08A%20City%20in%20a%20Garden.aspx.

Lau, Justine. 2008. "Buffett Buys BYD Stake." *Financial Times*. September 30. Accessed April 7, 2014. http://www.ft.com/intl/cms/s/0/235c9890-8de5-11dd-8089-0000779fd18c.html?siteedition=intl#ixzz2ZI7mmh2v.

Lawrence Berkeley Laboratory. 2012. "Berkeley Lab to Lead a U.S.-India Clean Energy Research Center." Press release. April 13. Accessed April 1, 2014. http://newscenter.lbl.gov/feature-stories/2012/04/13/berkeley-lab-to-lead-a-u-s-india-clean-energy-research-center/.

Lawson, Sam, and Larry MacFaul. 2010. *Illegal Logging and Related Trade: Indicators of the Global Response*. London: Chatham House.

Laylin, Tafline, 2013. "Inhabitat Tours Abu Dhabi's Masdar City." July 7. Accessed September 1, 2014. http://inhabitat.com/exclusive-new-photos-plus-energy-masdar-city-in-abu-dhabi/masdar-city-13/?extend=1.

Lechtenberg, Marie. 2012. "Top 20 Global Cement Companies." *Global Cement Magazine*. December 17. Accessed March 12, 2014. http://www.globalcement.com/magazine/articles/741-top-20-global-cement-companies.

Lee Kuan Yew. 1998. *The Singapore Story: Memoirs of Lee Kuan Yew*. Singapore: Straits Times Press.

——. 2006. *From Third World to First: The Singapore Story: 1965–2000*. Singapore: Singapore Press, Marshall Cavendish Editions.

——. 2008. "Minister Mentor Lee Kuan Yew's Dialogue at Singapore Energy Conference." Singapore Energy Conference. November 4. Accessed August 13, 2013. http://www.news.gov.sg/public/sgpc/en/media_releases/agencies/mica/transcript/T-20081105-1.html.

Lee Poh Onn. 2003. *The Water Issue Between and Singapore and Malaysia: No Solution in Sight?* Singapore: Institute of Southeast Asian Studies.

——. 2005. "Water Management Issues in Singapore." Paper Presented at Water in Mainland Southeast Asia. Siem Reap, Cambodia. November 29–December 2.

Lewis, Joanna I. 2013. *Green Innovation in China: China's Wind Power Industry and the Global Transition to a Low-Carbon Economy*. New York: Columbia University Press.

Lim, Richard. 1993. *Tough Men, Bold Visions: The Story of Keppel*. Singapore: Keppel Corp.

Lincoln, Abraham. 1953. "First Lecture on Discoveries and Inventions, 1858." *Collected Works of Abraham Lincoln. Volume* 2. New Brunswick, NJ Rutgers University Press.

Liu, Abby. 2013. "Skip the Cigarettes, Breathe in Beijing's Air." Global Voices. Accessed January 26, 2014. http://globalvoicesonline.org/2013/03/13/china-beijings-free-cigarettes-and-shanghais-pork-soup/.

Liu, Coco. 2011. "China's City of the Future Rises on a Wasteland." ClimateWire (via *New York Times*). September 28. Accessed March 10, 2014. http://www.nytimes.com/cwire/2011/09/28/28climatewire-chinas-city-of-the-future-rises-on-a-wastela-76934.html?pagewanted=all.

Ma, Wayne. 2013. "Beijing Pollution Hits Highs." *Wall Street Journal*. January 14. Accessed April 28, 2014. http://online.wsj.com/news/articles/SB10001424127887324235104578239142337079994.

MacMurray, Hannah. 2011. "Interview with Chetan Maini." GreenCarDesign. December 14. Accessed July 17, 2013. http://www.greencardesign.com/site/interviews/interview-chetan-maini-2011.

Maddison, Angus. 1995. *Monitoring the World Economy*. Paris: Development Center of the Organization for Economic Co-Operation and Development.

Mahindra Reva Electric Vehicles. 2013. Asset Light Process—Capital Investments 10x Lower than Traditional Automotive Product Development Cycles (unpublished company document).

Mahindra Rise. 2011. "An Electrifying Thinker: Exclusive Interview with Electric Vehicle Pioneer Chetan Maini." September 12. Accessed July 17, 2013. http://rise.mahindra.com/interview-with-electric-vehicle-pioneer-chetan-maini/.

Manila Water. 2010. 2009 Annual Report. Quezon City: Manila Water Company.

——. 2010. 2009 Sustainability Report. Quezon City: Manila Water Company.

——. 2011. 2010 Annual Report. Quezon City: Manila Water Company.

——. 2011. 2010 Sustainability Report. Quezon City: Manila Water Company.

——. 2012. 2011 Annual Report. Quezon City: Manila Water Company.

——. 2012. 2011 Sustainability Report. Quezon City: Manila Water Company.

——. 2013. 2012 Annual Report. Quezon City: Manila Water Company.

——. 2013. 2012 Sustainability Report. Quezon City: Manila Water Company.

——. "Manila Water Shares Best Practices in Singapore International Water Week." Press release. http://www.manilawater.com/investor/News%20and%20Updates/Pages/ReadNewsItem.aspx?ItemID=4.

Margono, Belinda Arunarwati, Peter V. Potapov, Svetlana Turubanova, Fred Stolle, and Matthew C. Hansen. 2014. Nature Climate Change. Volume 4. August. Published online June 29. "Primary Forest Cover Loss in Indonesia Over 200–2012." Accessed August 30, 2014. http://www.nature.com/nclimate/journal/vaop/ncurrent/full/nclimate2277.html.

Marks, Robert B. 2012. *China: Its Environment and History*. Lanham, MD: Rowman & Littlefield.

Marquis, Christopher, Honhyu Zhang, and Lixuan Zhou. 2013. "China's Quest to Adopt Electric Vehicles." *Stanford Social Innovation Review*. Accessed May 23, 2014. http://www.hbs.edu/faculty/Publication%20Files/Electric%20Vehicles_89176bc1–1aee-4c6e-829f-bd426beaf5d3.pdf.

Masdar City. "Masdar City: Frequently Asked Questions." Accessed August 10, 2013. http://masdarcity.ae/en/110/frequently-asked-questions.

Masson, Gaetan, Marie Latour, Manoel Rekinger, Ioannas-Thomas Theologitis, and Myrto Papoutsi. 2013. "Global Market Outlook for Photovoltaics 2013–2017." European Photovoltaic Industry Association. May. Accessed March 8, 2014. http://www.epia.org/fileadmin/user_upload/Publications/GMO_2013_-_Final_PDF.pdf.

Mattoo, Aaditya, and Arvind Subramanian. 2013. *Greenprint: A New Approach to Cooperation on Climate Change*. Washington, DC: Center for Global Development.

McElroy, Michael B., Xi Lu, Chris P. Nielsen, and Yuxuan Wang. 2009. "Potential for Wind-Generated Electricity in China." *Science* 325, no. 5946 (September 11):

1378–1380. doi:10.1126/science.1175706. Accessed May 28, 2014. http://www.sciencemag.org/content/325/5946/1378.full.html.

MacKinnon, Ian. 2007. *The Guardian.* "Four-Meter Floodwaters Displace 340,000 in Jakarta." February 5, 2007. Accessed August 30, 2014. http://www.theguardian.com/world/2007/feb/05/weather.indonesia.

Metropolitan Transportation Authority. "The MTA Network, Public Transportation for the New York Region." Accessed June 19, 2013. http://web.mta.info/mta/network.htm.

Miller, Tom. 2012. *China's Urban Billion: The Story Behind the Biggest Migration in Human History.* London: Zed Books.

Ministry of Economy, Trade and Industry of Japan. "Present Status and Promotion Measures for Renewable Energy in Japan: Importance of Renewable Energy." Accessed March 1, 2014. http://www.meti.go.jp/english/policy/energy_environment/renewable/ref1001.html.

Mitsubishi Electric Corporation. 2011. "Mitsubishi Electric to Install World's Fastest Elevators in Shanghai Tower." Press release. September 28. Accessed March 9, 2014. http://www.mitsubishielectric.com/news/2011/0928.html.

Mo, Kevin. 2009. "From Grey to Green: Make China's Rapid Urbanization Sustainable." Presentation at Natural Resources Defense Council Side Event During COP 15. Copenhagen. December 8.

MTR Corporation. 2010. Annual Report 2009. Accessed August 30, 2014. http://www.mtr.com.hk/eng/investrelation/2009frpt_e/EMTRAR2009F.pdf.

——. 2010. Sustainability Report 2009. Accessed August 30, 2014. http://www.mtr.com.hk/eng/sustainability/2009rpt/20Final/pdf/MTRSR09.pdf.

——. 2011. Annual Report 2010. Accessed August 30, 2014. http://www.mtr.com.hk/eng/investrelation/2010frpt_e/EMTRAR2010F.pdf.

——. 2011. Sustainability Report 2010. Accessed August 30, 2014. http://www.mtr.com.hk/eng/sustainability/2010rpt/sustainabilityreport2010.pdf.

——. 2012. Annual Report 2011. Accessed August 30, 2014. http://mtr.com.hk/eng/investrelation/2011frpt_e/EMTRAR2011F.pdf.

——. 2012. Sustainability Report 2011. Hong Kong. MTR Corporation. Accessed May 30, 2014. http://www.mtr.com.hk/eng/sustainability/2011rpt/sustainabilityreport2011.pdf.

——. 2013. Annual Report 2012. Accessed August 30, 2014. http://www.mtr.com.hk/eng/investrelation/2012frpt_e/EMTRAR2012F.pdf.

——. 2013. Sustainability Report 2012. Hong Kong. MTR Corporation. http://www.mtr.com.hk/eng/sustainability/2012rpt/files/sustainabilityreport2012.pdf. Accessed October 5, 2014.

——. 2014. Annual Report 2013. Accessed May 23, 2014. http://www.mtr.com.hk/eng/investrelation/2013frpt_e/EMTRAR2013F.pdf.

——. 2014. Express Rail Link. Accessed August 30, 2014. http://www.expressraillink.hk/en/home/.

Myrdal, Gunnar. 1968. *The Asian Drama: An Inquiry into the Poverty of Nations.* London: Allan Lane, The Penguin Press.

National Disaster Risk Reduction and Management Council. 2014. "SitRep No. 92 Effects of Typhoon 'Yolanda' (Haiyan)." January 14. Accessed August 24, 2014. http://reliefweb.int/sites/reliefweb.int/files/resources/NDRRMC%20Update%20re%20Sit%20Rep%2092%20Effects%20of%20%20TY%20%20YOLANDA.pdf.

National Geographic. 2012. "2012 Greendex Report." July 12. Accessed September 22, 2012. http://environment.nationalgeographic.com/environment/greendex/.

Japan, Government of. National Policy Unit. 2012. "Options for Energy and the Environment: The Energy and Environment Council Decision on June 29, 2012." Accessed March 8, 2014. https://s3-ap-northeast-1.amazonaws.com/sentakushi01/public/pdf/Outline_English.pdf.

National Renewable Energy Laboratory. 2013. "NREL Releases New Roadmap to Reducing Solar PV 'Soft Costs' by 2020." Press release. September 25. Accessed October 6, 2013. http://www.nrel.gov/news/press/2013/3301.html.

Nellemann, C., ed. 2012. *Green Carbon, Black Trade: Illegal Logging, Tax Fraud and Laundering in the World's Tropical Forests: A Rapid Response Assessment*. Arendal, Norway: UNEP/GRID-Arendal. http://www.unep.org/publications/contents/pub_details_search.asp?ID=6276.

Nestlé. 2010. "Growth in Vegetable Oils Usage for Bio-fuels Production." Media statement. April. Accessed June 7, 2014. http://www.Nestlé.com/asset-library/Documents/Media/Statements/oil-usage-for-biofuel-production.pdf.

——. 2012. "Nestlé Committed to Traceable Sustainable Palm Oil to Ensure No-deforestation." Media statement. October 30. Accessed June 7, 2014. http://www.Nestlé.com/Media/Statements/Pages/Update-on-deforestation-and-palm-oil.aspx#.UMANd4OTyYg.

Nestlé. 2013. "Progress Report Palm Oil." Autumn. Accessed June 7, 2014. http://www.nestle.com/asset-library/documents/creating-shared-value/responsible-sourcing/progress-report-palm-oil-autumn-2013.pdf.

Newton, James. 1987. *Uncommon Friends: Life with Thomas Edison, Henry Ford, Harvey Firestone, Alexis Carrel and Charles Lindbergh*. San Diego, CA: Harcourt Brace Jovanovich.

New York City Rent Guidelines Board. "2010 Housing Supply Report." Accessed August 14, 2013. http://www.housingnyc.com/downloads/research/pdf_reports/10HSR.pdf.

Ng, Eric. 2014 "Asia Pacific Resources eyes IPOs for three of its clean energy investments." *South China Morning Post*, October 16, 2014. http://www.scmp.com/business/money/markets-investing/article/1617151/asia-pacific-resources-eyes-ipos-three-its-clean.

Norway—The Official Site in Indonesia. 2010. "Norway-Indonesia REDD+ Partnership-Frequently Asked Questions." May 31. Accessed June 7, 2014. http://www.norway.or.id/Norway_in_Indonesia/Environment/-FAQ-Norway-Indonesia-REDD-Partnership-/#.UoNZqaiSzPw.

O'Dell, John. 2011. "New Investor Refocuses Boston Power on China." Edmunds AutoObserver. September 30. Accessed April 4, 2014. http://www.edmunds.com/autoobserver-archive/2011/09/new-investor-refocuses-boston-power-on-china.html.

Ogiso, Satoshi. 2013. "Toyota's Hybrid Technology" (unpublished company document).

ORIX Corporation. 2013. "ORIX Acquires Robeco." Press release. February 19. Accessed October 5. 2014. http://www.orix.co.jp/grp/en/pdf/news/130219_ORIXE1.pdf.

——. 2013. "Japan's Largest Condo Building Implementing Bulk Electric Power Purchasing to Use Green Power." Press release. May 28. Accessed October 5, 2013. http://www.orix.co.jp/grp/en/news/2012/120528_OepE.html.

——. 2013. "Construction to Start on Mega-Solar Project at Tokachi Speedway." Press release. July 5. Accessed October 5, 2014. http://www.orix.co.jp/grp/en/news/2013/130705_ORIXE.html.

——. 2013. "Construction Starts on a Mega-Solar Project with a Maximum Output of 11.7 MW in Omuta City, Fukuoka Prefecture." Press release. September 4. Accessed March 8, 2014. http://www.orix.co.jp/grp/en/news/2013/130904_ORIXE.html.

——. 2013. "Toshiba and ORIX to Develop Geothermal Power Generation Business in Nakao, Okuhida Onsen, Gifu Prefecture." Press release. November 19. Accessed June 7, 2014. http://www.toshiba.co.jp/about/press/2013_11/pr1901.htm.

Parkinson, Giles. 2012. "Interview: First Solar CEO James Hughes." RenewEconomy.com. December 13. Accessed March 8, 2014. http://reneweconomy.com.au/2012/interview-first-solar-ceo-james-hughes-72086.

Pearce, Fred. 2011. "The Triumph of King Coal: Hardening Our Coal Addiction." Yale Environment 360. October 31. Accessed March 5, 2014. http://e360.yale.edu/feature/the_triumph_of_king_coal_hardening_our_coal_addiction/2458/.

Peleg, Barchi. 2007 (rev. 2013). "Case GS-48: Esquel Group: A Vertically Integrated Apparel Manufacturer." Stanford, CA: Stanford Graduate School of Business.

Pennington, Matthew. 2013. "Singapore PM Draws Laughs in US Speech." Associated Press. April 3. Accessed January 29, 2014. http://news.yahoo.com/singapore-pm-draws-laughs-us-speech-111914557—politics.html.

Pernick, Ron, and Clint Wilder. 2008. *The Clean Tech Revolution: Discover the Top Trends, Technologies and Companies to Watch*. New York: Collins Business.

Peters, Siobhan, et al. 2006. *The Stern Review: Economics of Climate Change*. London: Government of Britain. Accessed May 28, 2014. http://mudancasclimaticas.cptec.inpe.br/~rmclima/pdfs/destaques/sternreview_report_complete.pdf.

Pettis, Michael. 2013. "Monetary Policy Under Financial Repression." December 20. Accessed February 4, 2014. http://blog.mpettis.com/2013/12/monetary-policy-under-financial-repression/.

Philip, Bruno. 2013. "Jakarta Faces Up to a High Flood-Risk Future." *The Guardian*. February 5. Accessed May 28, 2014. http://www.theguardian.com/world/2013/feb/05/jakarta-floods-rising-sea-levels.

Popular Science. 2013. "New-topias." June. Accessed August 10, 2013. http://www.songdo.com/Uploads/FileManager/Popular%20Science%206%202013.pdf.

Powell, Bill, and Charlie Zhu. 2013. "Special Report: The Rise and Fall of China's Sun King." Reuters. May 18. Accessed October 9, 2013. http://www.reuters.com/article/2013/05/19/us-suntech-shi-specialreport-idUSBRE94I00220130519.

PUB Singapore. 2011. *Innovation in Water*. Singapore: Public Utilities Board, June.

——. 2012. *Innovation in Water*. Singapore: Public Utilities Board, March.

——. 2012. *Innovation in Water*. Vol. 3. Singapore: Public Utilities Board, July. Retrieved February 10, 2013. http://www.pub.gov.sg/mpublications/Lists/WaterInnovationPublication/Attachments/9/InnovationWater_vol3.pdf

PwC. 2012. "Photovoltaic Sustainable Growth Index." October. Accessed March 6, 2014. http://www.pwc.com/en_US/us/technology/assets/pwc-pv-sustainable-growth-index.pdf.

Qiao Liming. 2012. "China's Wind Development: Experiences Gained and Lessons Learnt." Presentation at IRENA Workshop. Copenhagen. April. Accessed May 28, 2014. https://www.irena.org/DocumentDownloads/events/CopenhagenApril2012/5_Liming_Qiao.pdf.

Reuters. 2014. "California Electric Grid Sets Solar Generation Record." March 10. Accessed May 26, 2014. http://www.reuters.com/article/2014/03/10/utilities-california-solar-idUSL2N0M724F20140310.

——. "Chuanfu, Wang." Accessed April 7, 2014. http://in.reuters.com/finance/stocks/officerProfile?symbol=RENN.N&officerId=1742100.

——. "Nestlé Sets Social, Environmental Targets for 2020." Accessed April 4, 2013. http://www.reuters.com/article/2013/03/13/Nestlé-idUSL6N0C563120130313.

Rieley, J. O., and S. E. Page. 2005. "Wise Use of Tropical Peatlands: Focus on Southeast Asia." Wageningen, The Netherlands: ALTERRA—Wageningen University and Research Center and the EU INCO—STRAPEAT and RESTORPEAT Partnerships.

Rivera, Virgilio C., Jr. 2014. *Tap Secrets: The Manila Water Story*. July. Manila: Asian Development Bank and Manila Water. Accessed August 30, 2014. http://www.adb.org/sites/default/files/pub/2014/tap-secrets.pdf.

Robins, Nick, Zoe Knight, and Wai-shin Chan. 2013. "2013: The Great Disconnect." HSBC Global Research. January 7. Accessed June 6, 2014. https://www.research.hsbc.com/midas/Res/RDV?ao=20&key=NHoQRor6Qo&n=355330.PDF.

Rosa, Marc. 2013. "Global Solar Capacity Tops 100 Gigawatts on Asian Markets." Bloomberg Sustainability. February 11. Accessed March 8, 2013. http://www.bloomberg.com/news/2013-02-11/global-solar-capacity-tops-100-gigawatts-on-asian-markets.html.

Rosenthal, Elizabeth. 2012. "British Soil Is Battlefield over Peat, for Bogs' Sake." *New York Times*. October 6. Accessed June 7, 2014. http://www.nytimes.com/2012/10/07/science/earth/british-gardeners-battle-over-peat-for-bogs-sake.html.

Sankhe, Shirish, et al. 2010. "India's Urban Awakening: Building Inclusive Cities, Sustaining Economic Growth." McKinsey Global Institute. April. Accessed August 31, 2014. http://www.mckinsey.com/insights/urbanization/urban_awakening_in_india.

SCG (Siam Cement Group). 2012. Sustainability Report 2011. Bangkok: SCG.

SCG (Siam Cement Group). 2013. Sustainability Report 2012. Bangkok: SCG.

SCG (Siam Cement Group). 2014. Sustainability Report 2013. Bangkok: SCG.

Schlafly, Phyllis. 2013. "Obama's Giveaway to the Communists." WND. January 14. Retrieved April 4, 2014. http://www.wnd.com/2013/01/obamas-giveaway-to-the-communists/.

Self, Dan, and Jessie Morris. 2013. "Lowering the Cost of Solar PV: Soft Costs with Hard Challenges (Part 1 of 2)." Rocky Mountain Institute. September 25. Accessed

October 6, 2013. http://blog.rmi.org/blog_2013_09_25_lowering_the_cost_of_solar_PV_part_one.

Sembcorp. 2013. "Quick Facts." Updated March 11. Accessed February 11, 2014. http://www.sembcorp.com/en/about-quick-facts.aspx.

Shi Lishan. 2010. *Three Decades of Wind Power in China*. Beijing: Chinese Renewable Energy Industries Association.

SinaEnglish. 2013. "191 More Dead Pigs in Huangpu." March 3. Accessed January 26, 2014. http://english.sina.com/china/2013/0324/574949.html.

Singapore, Government of. "BCA's 2nd Green Building Masterplan—80% of Buildings Will Be Green by 2030." Building and Construction Authority. Press release. Accessed February 7, 2014. http://www.bca.gov.sg/newsroom/others/pr270409.pdf.

——. 2nd Green Masterplan. Building and Construction Authority. Accessed May 29, 2014. http://www.bca.gov.sg/greenMark/others/gbmp2.pdf.

——. Building and Construction Authority. [no date] "Singapore: Leading the Way for Green Buildings in the Tropics." Accessed August 29, 2014. https://www.bca.gov.sg/greenmark/others/sg_green_buildings_tropics.pdf.

——. "Tapping into Singapore's Pool of Resources." Environment & Water Industry Programme Office. http://edb.gov.sg/content/dam/edb/en/industries/Environment%20and%20Water/downloads/Global%20Hydrohub%20Brochure.pdf.

——. "Our City in a Garden." National Parks Board. Accessed August 4, 2013. http://www.nparks.gov.sg/ciag/.

——. "Gardens by the Bay: Our Legacy in Green." No date [2011]. Singapore: National Parks Board. Accessed August 29, 2014. http://www.gardensbythebay.com.sg/content/dam/gbb/en/images/get-involved/NP11_0082_GardenBrochure_220x307_R10.pdf.

Singapore Business Review. 2013. "Chart of the Day: MRT Ridership Could Grow 89% to 1.3b in 2030." February 15. Accessed June 7, 2014. http://sbr.com.sg/economy/news/chart-day-mrt-ridership-could-grow-89–13b-in-2030.

Singh, Shreyasi. 2010. "How I Did It: Chetan Maini." India Inc. March. Accessed July 17, 2013. http://www.growthinstitute.in/emagazine/mar10/resource-center.html.

Sino-Singapore Tianjin Eco-city. Accessed June 6, 2014. http://www.tianjinecocity.gov.sg/.

Skidmore, Owings & Merrill. 2013. "Nation Building by Design: SOM in China." Accessed August 3, 2013. http://www.som.com/ideas/publications/nation_building_by_design_som_in_china.

Smil, Vaclav. 1993. *China's Environmental Crisis: An Inquiry into the Limits of National Development*. Armonk, NY: M. E. Sharpe.

Songdo International Business District. 2012. "Korea's Songdo International Business District—One of Asia's Largest Green Developments—Surpasses Milestone of 13 Million Feet of LEED Certified Space." Press release. June 27. Accessed March 13, 2014. http://www.songdo.com/songdo-international-business-district/news/press-releases.aspx/.

Sovacool, Benjamin K. 2014. "Environmental Issues, Climate Changes, and Energy Security in Developing Asia." June. Page 3. Manila: Asian Development Bank. ADB Economics Working Paper Series No. 399. Accessed August 30, 2014.

Steiner, Achim. 2010. "Korea's Green New Deal." In *Korea 2020: Imagining the Next Decade*. Seoul: Random House Korea.

Straits Times. 2013. "ESM Goh: 'The Singapore Child Is Being Suffocated.'" June 21. Accessed May 6, 2014. http://www.straitstimes.com/breaking-news/singapore/story/esm-goh-the-singapore-child-being-suffocated-20130621.

——. 2013. "PM Lee's Press Conference on Haze, June 20." June 21.

Sudworth, John. 2013. "China Pulls Nearly 6,000 Dead Pigs from Shanghai River." BBC News. March 13. Accessed January 26, 2014. http://www.bbc.co.uk/news/world-asia-china-21766377.

Sumalee, Agachai. 2014. "Rail + Property + Pedestrian Model (RPP): Case Study of Hong Kong and Implication for Thailand." Department of Civil and Structural Engineering, Hong Kong Polytechnic University. Accessed May 29, 2014. http://www.atransociety.com/2014/pdf/pdfSymposium2010/download/ppt_day_1/(3)%20Dr.%20Agachai%20Sumalee.pdf.

Suntech Power Holdings. 2006. FY 05 Q3 Form 20-F for the Period Ending December 31, 2005. Filed April 27, 2006. Accessed May 26, 2014. https://www.sec.gov/Archives/edgar/data/1342803/000114554906000550/h00483e20vf.htm.

——. 2007. FY06 Form 20-F for the Period Ending December 31, 2006. Filed June 18, 2007. Accessed March 5, 2014. http://www.sec.gov/Archives/edgar/data/1342803/000114554907001069/h01290e20vf.htm.

——. 2011. 2010 Form 20-F for the Period Ending December 31, 2010. Filed June 6, 2011. Accessed March 5, 2014. http://www.sec.gov/Archives/edgar/data/1342803/000095012311047433/h04443e20vf.htm.

——. 2012. 2011 Form 20-F for the Period Ending December 31, 2011. Filed April 27, 2012. Accessed March 5, 2014. http://www.sec.gov/Archives/edgar/data/1342803/000114554907001069/h01290e20vf.htm.

——. 2012. 2011 Annual Report. April 30. http://ir.suntech-power.com/phoenix.zhtml?c=192654&p=irol-reportsAnnual.

Swire Pacific. 2013. Annual Report 2012. Hong Kong: Swire Pacific. http://www.swirepacific.com/en/ir/reports/swirepacificAR2013/pdf/en/Swire_AR13_e_140401.pdf

Swire Pacific. 2014. Annual Report 2013. Hong Kong: Swire Pacific. http://www.swirepacific.com/en/ir/reports/swirepacificAR2012/pdf/en/SwireAR12_E.pdf

Tahija, Jean. 1998. *An Unconventional Woman*. Ringwood, Victoria: Penguin Books Australia.

Tahija, Julius. 1995. *Horizon Beyond*. Singapore: Times Editions.

Takada, Aya, and Chisaki Watanabe. 2014. "Solar Farmers in Japan to Harvest Electricity with Crops." Bloomberg. May 27. Accessed May 28, 2014. http://www.bloomberg.com/news/2014–05–26/solar-farmers-in-japan-to-harvest-electricity-with-crops.html.

Taraporevala, Sooni. 2004. *Parsis: The Zoroastrians of India: A Photographic Journey, 1980–2004*. Mumbai: Good Books.

Taylor, Alex III. 2006. "The Birth of the Prius." *Fortune*. February 24. Accessed February 8, 2014. http://money.cnn.com/magazines/fortune/fortune_archive/2006/03/06/8370702/.

Thomas, Mike. 2014. "The Opportunity for Coal in the Context of Natural Gas." Paper Presented at the Coaltrans China Conference. Shanghai. April 10. Hong Kong: The Lantau Group.

Time. 1969. "America's Sewage System and the Price of Optimism." August 1.

Toms, Manu P. 2012. "Mahindra Reva Targets 30,000 Electric Cars a Year by 2015–16." *Hindustan Times*. August 22. Accessed July 17, 2013. http://www.hindustantimes.com/Autos/HTAuto-TopStories/Mahindra-Reva-targets-30–000-electric-cars-a-year-by-2015–16/Article1–917825.aspx.

Tortajada, Cecilia, Yugal Joshi, and Asit K. Biswas. 2013. *The Singapore Water Story: Sustainable Development in an Urban City-State*. Milton Park: Routledge.

Toshiba. 2013. "Toshiba and ORIX to Develop Geothermal Power Generation Business in Nakao, Okuhida Onsen, Gifu Prefecture." Press release. November 19. Accessed March 8, 2014. http://www.toshiba.co.jp/about/press/2013_11/pr1901.htm.

Towers Watson. 2014. "Manila Water Company: 10 Questions for Ruel Maranan, Group Director, Corporate Human Resources." January. Accessed January 29, 2014. http://www.towerswatson.com/en/Insights/Newsletters/Global/strategy-at-work/2014/10-questions-for-ruel-maranan.

Toyota. 2013. "Toyota to Launch 'New Era' of High MPG Hybrids, Expand Its Global Hybrid Rollout." Press release. August 28. Accessed May 29, 2014. http://toyotanews.pressroom.toyota.com/releases/2013+toyota+global+hybrid+rollout.htm.

——. 2014. "Worldwide Sales of Toyota Hybrids Top 7 Million Units." Press release. October 14. Accessed October 22, 2014. http://newsroom.toyota.co.jp/en/detail/4069183/

Transport for London. "Creating a New Ultra Low Emission Discount." Accessed July 17, 2013. https://consultations.tfl.gov.uk/roads/5503a5b6.

——. "Key Facts." Accessed June 19, 2013. http://www.tfl.gov.uk/corporate/modesoftransport/londonunderground/1608.aspx.

Unilever. 2014. "Our Sustainable Sourcing Journey," Accessed July 9, 2014. http://www.unilever.com/sustainable-living-2014/reducing-environmental-impact/sustainable-sourcing/sustainable-palm-oil/

United Nations Department of Economic and Social Affairs. 2013. "World Urbanization Prospects, the 2011 Revision." October 7. Accessed March 8, 2014. http://esa.un.org/unup/.

United Nations Development Program. 2011. "India Global Environment Facility Project Document." Accessed May 29, 2014. http://www.undp.org/content/dam/india/docs/energy_efficiency_improvements_in_commercial_buildings_project_document.pdf.

United Nations Economic and Social Commission for Asia and the Pacific. 2009. "Sustainable Agriculture and Food Security in Asia and the Pacific." http://climate-l.iisd.org/news/unescap-releases-study-on-sustainable-agriculture-and-food-security-in-asia-and-the-pacific/.

——. *Statistical Yearbook for Asia and the Pacific 2013*. Bangkok: United Nations publication. See http://www.unescap.org/stat/data/syb2013/ESCAP-syb2013.pdf.

United Nations Environment Program, Sustainable Buildings & Climate Initiative. 2009. "Buildings and Climate Change: Summary for Decision-Makers. Accessed August 29, 2014. http://www.unep.org/sbci/pdfs/sbci-bccsummary.pdf

U.S. Energy Information Administration. 2014. "China is Now the World's Largest Importer of Petroleum and Other Fuels." March 24. Accessed August 24, 2014. http://www.eia.gov/todayinenergy/detail.cfm?id=15531.

U. S. Energy Information Administration. "Energy Intensity—Total Primary Energy Consumption per Dollar of GDP (Btu per Year 2005 U.S. Dollars (Market Exchange Rates))." Accessed May 26, 2014. http://www.eia.gov/cfapps/ipdbproject/iedindex3.cfm?tid=92&pid=46&aid=2.

——. Frequently Asked Questions: How Much Electricity is Used for Lighting in the United States?" Accessed August 29, 2014. http://www.eia.gov/tools/faqs/faq.cfm?id=99&t=3.

U.S. Environmental Protection Agency. "Global Greenhouse Gas Emissions Data." Accessed August 30, 2014. http://www.epa.gov/climatechange/ghgemissions/global.html.

U.S. Trade Representative. 2014. "The People's Republic of China: U.S.-China Trade Facts." Accessed August 30, 2014. http://www.ustr.gov/countries-regions/china-mongolia-taiwan/peoples-republic-china.

Urban Land Institute and Center for Liveable Cities. 2013. *10 Principles for Liveable High-Density Cities: Lessons from Singapore*. Singapore: Urban Land Institute and Center for Liveable Cities.

Vogel, Ezra P. 2011. *Deng Xiaoping and the Transformation of China*. Cambridge: Belknap Press.

Wang Hui. "China Plans Five-Year Leap Forward of Railway Development." China Features. Accessed April 9. 2014. http://bg.chineseembassy.org/eng/dtxw/t274660.htm.

Watts, Jonathan. 2009. "China Teams Up with Singapore to Build Huge Eco-city." *The Guardian*. June 4. Accessed March 10, 2014. http://www.guardian.co.uk/world/2009/jun/04/china-singapore-tianjin-eco-city.

——. 2010. *When a Billion Chinese Jump: How China Will Save Mankind—or Destroy It*. London: Faber and Faber.

Wen Hong, Madelaine Steller Chiang, Ruth A. Shapiro, and Mark L. Clifford. 2007. *Building Energy Efficiency: Why Green Buildings Are Key to Asia's Future*. Hong Kong: Inkstone Books.

Wheelan, Hugh. 2013. "Paris: Burgeoning Green Building Regs to Impact Property Holdings Value Says €7.5 Trillion Investor, Group Policy Makers Target Heavy CO_2 Emitting Building Sector." Responsible Investor. March 25. Accessed June 7, 2013. http://www.iigcc.org/news,-press-and-events.

Wiser, Ryan, and Matt Bolinger. 2013. "2012 Annual Wind Market Report." U.S. Department of Energy. August. Accessed March 8, 2014. http://www.windpoweringamerica.gov/pdfs/2012_annual_wind_market_report.pdf.

Woetzel, Jonathan. "China's Cities in the Sky." McKinsey & Co. Accessed February 9, 2014. http://voices.mckinseyonsociety.com/chinas-cities-in-the-sky/#sthash.ZKqeV2G7.dpuf.

Woetzel, Jonathan, et al. 2009. "Preparing for China's Urban Billion." McKinsey & Co. February. Accessed May 28, 2014. http://www.mckinsey.com/insights/urbanization/preparing_for_urban_billion_in_china.

Wong, Julian. 2008. "Creating a Better Life: A Closer Look at the Sino-Singapore Tianjin Eco-city Project." Green Leap Forward. November 16. Accessed March 10, 2014. http://greenleapforward.com/2008/11/16/creating-a-better-life-a-closer-look-at-the-sino-singapore-tianjin-eco-city-project/.

Wong, Sue-lin, and Clare Pennington. 2013. "Steep Challenges for a Chinese Eco-city." *New York Times*. February 13. Accessed March 9, 2014. http://green.blogs.nytimes.com/2013/02/13/steep-challenges-for-a-chinese-eco-city/.

Woody, Tom. 2014. "The Chart That Shows Why China Is Desperate to Switch to Electric Cars." *The Atlantic*. May 19. Accessed May 23, 2014. http://www.theatlantic.com/technology/archive/2014/05/the-chart-that-shows-why-china-is-desperate-to-switch-to-electric-cars/371153/.

World Bank. 2010. "Private Concessions: The Manila Water Experience." March. Accessed May 28, 2014. http://siteresources.worldbank.org/NEWS/Resources/ManilaWaterProject3–31–10.pdf.

——. 2010. "Chinese Eco-city Project Gets Boost from Global Environment Facility." July 22. Accessed March 10, 2014. http://www.worldbank.org/en/news/press-release/2010/07/22/chinese-eco-city-project-gets-boost-global-environment-facility.

——. 2012. "An Update to the World Bank's Estimates of Consumption Poverty in the Developing World." February 29. Accessed March 5, 2014. http://siteresources.worldbank.org/INTPOVCALNET/Resources/Global_Poverty_Update_2012_02–29–12.pdf.

——. Data: GDP Growth (Annual %). Accessed August 29, 2014. http://data.worldbank.org/indicator/NY.GDP.MKTP.KD.ZG.

——. "Water Pollution, Textile Industry (% of Total BOD Emissions)." World Development Indicators. Accessed May 28, 2014. http://data.worldbank.org/indicator/EE.BOD.TXTL.ZS?page=1.

World Bank and PRTM Management Consultants. 2011. "The China New Energy Vehicles Program: Challenges and Opportunities." April. Accessed May 30, 2014. http://siteresources.worldbank.org/EXTNEWSCHINESE/Resources/3196537–1202098669693/EV_Report_en.pdf.

World Future Energy Summit, 2012. "His Highness Sheikh Mohamed bin Zayed Launches International Water Summit." January 18. Accessed August 24, 2014. http://www.worldfutureenergysummit.com/Portal/news/18/1/2012/his-highness-sheikh-mohamed-bin-zayed-launches-international-water-summit.aspx

World Health Organization. 2014. "Air Quality Deteriorating in Many of the World's Cities." May 7. Accessed August 31, 2014. http://www.who.int/mediacentre/news/releases/2014/air-quality/en/.

——. 2014. Ambient (Outdoor) Air Pollution in Cities Database 2014. Accessed August 31, 2014. http://www.who.int/phe/health_topics/outdoorair/databases/cities/en/.

World Resources Institute. 2014. Climate Analysis Indicators Tool (CAIT 2.0). "Total GHG Emissions Including Land-Use Change and Forestry-2011" Accessed August 31, 2014. http://cait2.wri.org/wri/Country%20GHG%20Emissions?

indicator[]=Total%20GHG%20Emissions%20Excluding%20Land-Use%20Change%20and%20Forestry&indicator[]=Total%20GHG%20Emissions%20Including%20Land-Use%20Change%20and%20Forestry&year[]=2011&sortIdx=1&sortDir=desc&chartType=geo.

World Wildlife Fund. 2013. "2013 Palm Oil Buyers Scorecard." Accessed April 1, 2014. http://wwf.panda.org/what_we_do/footprint/agriculture/palm_oil/solutions/responsible_purchasing/palm_oil_buyers_scorecard_2013/.

Wu Wencong. 2012. "Pollution Blind Spot in the Textile Industry." *China Daily*. October 9. Accessed June 7, 2014. http://www.chinadaily.com.cn/china/2012–10/09/content_15802134.htm.

Xinhua. 2010. "Recharging China's Clean Energy Dream." October 2. Accessed March 8, 2014. http://news.xinhuanet.com/english2010/indepth/2010–10/02/c_13539987.htm.

Xinjiang Goldwind Science & Technology Co., Ltd. 2010. Global Offering [Shares to be listed on The Stock Exchange of Hong Kong]. Joint Global Coordinators: China International Capital Corporation (CICC) and Citibank. Accessed August 28, 2014. http://pg.jrj.com.cn/acc/CN_DISC%5CSTOCK_NT%5C2010%5C06%5C07%5C002202_ls_58035848.PDF.

Yoneda, Yuka. 2011. "Tianjin Eco City Is a Futuristic Green Landscape for 350,000 Residents." Inhabitat. January 10. Accessed March 12, 2014. http://inhabitat.com/tianjin-eco-city-is-a-futuristic-green-landscape-for-350000-residents/.

Yu Ran. 2014. "A Wealth of Chinese Billionaires Appears on Global Rich List." *China Daily USA*. February 26. Accessed May 8, 2014. http://usa.chinadaily.com.cn/epaper/2014–02/26/content_17307586.htm.

Yu Yongding. 2013. "Can China Wean Itself Off Its Addiction to Investment?" *South China Morning Post*. October 11. Accessed October 12, 2013. http://www.scmp.com/comment/insight-opinion/article/1328691/can-china-wean-itself-its-addiction-investment.

Zhang Yi. 2013. "Pig Pollution Shows Urgent Need to Ensure Water Quality." *Global Times*. March 3. Accessed January 26, 2014. http://www.globaltimes.cn/content/767692.shtml#.UuS283lm4y4.

Zöllter, Jürgen. 2011. "Mercedes Set to Launch New Electric Car in China" (WorldCrunch translation of *Die Welt* article). December 7. Accessed April 7, 2014. http://www.worldcrunch.com/business-finance/mercedes-set-to-launch-new-electric-car-in-china-.

Index